D1218303

The General Theory of Alternating Current Machines:

Application to Practical Problems

The General Theory of Alternating Current Machines:
Application to Practical Problems

BERNARD ADKINS
M.A., D.Sc. (Eng.), F.I.E.E.
Research Fellow, Imperial College, London SW7.

RONALD G. HARLEY
Pr.Eng., M.Sc. Eng., Ph.D., M.I.E.E.
Professor of Electrical Machines and Control,
University of Natal, Durban.

CHAPMAN AND HALL
LONDON

First published 1975
by Chapman and Hall Ltd.
11 New Fetter Lane, London EC4P 4EE.

Typeset by
Santype Ltd. (Coldtype Division)
Salisbury, Wiltshire

Printed and bound in Great Britain by
Redwood Burn Limited
Trowbridge & Esher

ISBN 0 412 12080 1

Distributed in the U.S.A.
by Halsted Press, a Division of
John Wiley & Sons, Inc., New York

Library of Congress Catalog Card Number 75-2107

Preface

The book on *The General Theory of Electrical Machines*, by B. Adkins, which was published in 1957, has been well received, as a manual containing the theories on which practical methods of calculating machine performance can be based, and as a text-book for advanced students. Since 1957, many important developments have taken place in the practical application of electrical machine theory. The most important single factor in the development has been the increasing availability of the digital computer, which was only beginning to be used in the solution of machine and power system problems in 1957. Since most of the recent development, particularly that with which the authors have been concerned, has related to a.c. machines, the present book, which is in other respects an up-to-date version of the earlier book, deals primarily with a.c. machines.

The second chapter on the primitive machine does deal to some extent with the d.c. machine, because the cross-field d.c. generator serves as an introduction to the two-axis theory and can be used to provide a simple explanation of some of the mathematical methods. The equations also apply directly to a.c. commutator machines.

The use of the word 'general' in the title has been criticized. It was never intended to imply that the treatment was *comprehensive* in the sense that every possible type of machine and problem was dealt with. The word is used in the sense that the theory *can* be applied to all types of machine and all conditions of operation. Strictly of course the theory does not apply exactly to any machine, but only to an idealized model, which is similar to the practical machine. The whole range of machines can be divided into three categories,

1. Those for which the theory can give an accurate prediction of its behaviour from design details.
2. Those where there are larger discrepancies but where the theory helps in obtaining an understanding of the problem.
3. Those for which the theory cannot be usefully employed.

The great majority of synchronous and induction machines in use at the present time fall into the first category, mainly because the design process aims at the elimination of harmonics and other factors which constitute the difference between the idealized and the practical machine.

Although the book is still mainly concerned with the two-axis theory and its development, the generalization has been extended to cover other reference frames, particularly that in which the variables are the actual phase currents in an a.c. armature. In recent years an increasing number of special machines, like linear motors, inductor alternators, reluctance motors or new types of change-pole motors, are being used. Sometimes the two-axis theory cannot be used or would be too inaccurate, but a theory based on the phase equations would be satisfactory. Without the computer a solution would usually not be practicable, but the equations can now be readily programmed for computation. Saturation and eddy current effects can often be handled within the framework of the theory, as explained in Chapter 10, but in some of the special machines the results would be too inaccurate. The authors nevertheless make no apology for continuing to use the word *general*, with the above qualifications.

In the interest of greater mathematical rigour, Laplace transforms are introduced when solutions of linear equations are derived. However, the Heaviside formulation, in which the symbol p is regarded as an operator, is used both in the initial statements of equations, linear and non-linear, and in the application of operational transfer functions and equivalent circuits.

Another change, which is a result of increasing use of computers, is that a short explanation is given in Chapter 5 of some of the modern control theories for which the equations are expressed in state variable form. The methods are particularly applicable to systems with multiple controls and to those for which the equations are non-linear. Some examples of their application to step-by-step computations are given in Chapter 9.

Some changes of notation have been made to conform to the new international specifications. The term *phasor* is used instead of *vector* for the complex number representing a sinusoidal quantity, while the word *vector* is used for a set of state-space variables.

Over a period of twenty years, the micro-machine equipment at Imperial College has been used by a team of research workers to study problems relating to a.c. machines and the external systems and devices associated with them. They have also had the opportunity of studying the results of tests made on large machines in factories or power stations. The programme of work at Imperial College is only one of many such activities round the world, but it has been found convenient to use the results of this programme, which are recorded in a number of I.E.E. papers, for many of the practical applications of the methods described in the book. As stated above, the theory is of general application, but it is only necessary to give a few selected illustrations. A bibliography giving references to all the work on the subject would be unduly long, but many other publications can be found from bibliographies of the listed papers.

For the theoretical derivations, the plan is adopted that a full explanation is given for the basic theory and the most important formulas. On the other hand, for the theory relating to many of the practical applications, for which the derivations are explained in the quoted references, only abbreviated explanations are given.

The book is intended for students and others who already have some knowledge of electrical machines, since it does not include any description of their construction. The mathematics required includes complex algebra, matrix algebra, Laplace transforms and a few other matters which are explained briefly in the text.

Contents

Introduction

The purpose of the book is to present a general theory of rotating electro-magnetic machines, applicable to all the normal types of machine and to all conditions of operation, and is consequently more fundamental and of wider application than the usual theories given in the standard textbooks. The theory applies to all machines in which alternate magnetic poles are formed round a cylindrical surface.

An analytical study of electrical machines consists of two parts:

1. Determination of the basic characteristics expressed by a number of quantities known as the machine *parameters*. In making calculations the values are often assumed to be constant, and the parameters are then referred to as *constants*.
2. Calculation from the constants of the performance of the machine under given external conditions.

The term 'theory of electrical machines', as interpreted in this book, refers only to the second part. The theory starts with an idealized machine, the properties of which are expressed by known constants, and provides a means of calculating its performance. For the purposes of the theory, the constants, which are essentially resistances and inductances, must be carefully defined, and the principles underlying their calculation clearly stated. Details of the methods of calculating them, with which the first of the above two parts is concerned, although very important for practical design work, form a separate subject.

In the usual textbook theories of electrical machines each type of machine is dealt with on its own merits with little reference to other types, and simple methods of analysis are developed by means of which the performance under specified conditions can be calculated. In these theories the main emphasis is on steady operation, and they lead to graphical or analytical methods of calculation. For a.c. machines, phasor diagrams are very widely used. The standard approach has the disadvantage that a completely fresh start has to be made when it is necessary to analyse a new type of machine or to deal with unbalanced or transient conditions.

It is interesting to survey the historical development of the theory of electrical machines. The early theories of a.c. machines were based on the phasor diagram and were worked out by geometrical constructions on a drawing-board. There followed a search for equivalent circuits, which finally led to the acceptance of a few standard circuits selected from a large number of possible ones. The next important development was the introduction of complex numbers in what was known as the 'symbolic method' or the 'j-method'. At that time, however, the algebraic method was not introduced in its own right, but only as an auxiliary process to assist in working out the phasor diagrams or equivalent circuits. The modern methods involve a new approach to the subject. In the modern theory the algebraic equations are accepted as the fundamental means of expression, and phasor diagrams and equivalent circuits become merely devices leading to alternative methods of solution applicable only to special cases. The use of equations is in line with the accepted methods of circuit theory, and leads to a general theory of electrical machines which embraces all types and all conditions of operation.

The fundamental set of equations is derived for an idealized two-pole machine, which is approximately equivalent to the actual machine, in accordance with certain well-defined assumptions. In general they are differential equations in which an applied voltage is equated to the sum of several component voltages which depend on the currents, or an applied torque is equated to the sum of component torques. In d.c. machines the equations relating the actual currents with the voltages and torque are usually in a convenient form for practical solution, but for most problems in a.c. machines the equations tend to be complicated and difficult to solve, although a solution is possible with a digital computer.

However, for the great majority of practical a.c. machines, including normal induction and synchronous machines, a great simplification is obtained by expressing the equations in a new reference frame and introducing certain fictitious currents and voltages which are different from but are related to the actual ones. The fictitious currents can have a physical meaning in that they can be considered to flow in fictitious windings acting along two axes at right angles, called the direct and quadrature axes. In this way a 'two-axis theory' of a.c. machines is developed, and the equations so obtained are found to correspond very closely to those of d.c. machines.

Because of its practical importance, a large part of the book is devoted to applications of the two-axis theory, particularly to synchronous and induction machine problems. The first step in its development was Blondel's 'two-reaction theory' of the steady-state operation of the salient-pole synchronous machine. The method was examined in detail by Doherty and Nickle, who published a series of five important papers [5]. A paper by West on 'The Cross-field Theory of Alternating Current Machines' [2] assumed without proof that a rotating cage winding is equivalent to a d.c. armature winding with two short-circuited pairs of brushes. A very valuable contribution to the subject was made by Park in a set of three papers [3, 4 and 6]. These papers not only develop the general two-axis equations of the synchronous machine, but they indicate how the equations can be applied to many important practical problems. Park's transformation provides the most important fundamental concept in the development of Kron's generalized theory, which was first published in a series of papers in The General Electric Review, and later in a book [8]. Many more recent books and papers on the subject are referred to in the bibliography.

After discussing some general matters in Chapter 1, the equations of the primitive machine and some simple applications are considered in Chapter 2. Chapters 3 and 4 develop the theory of the a.c. machine in terms both of the actual armature variables and of the two-axis variables. Chapters 5 and 6 set down some fundamental matters relating to methods of analysis and automatic control and Chapters 7 to 11 are devoted to the application of the two-axis theory to many practical problems arising in the application of synchronous and induction machines. Chapter 12 discusses problems relating to other less common types of machine, including some to which the two-axis method is not applicable. Four short appendices consider some special matters which arise in applying some of the well-known mathematical methods to machine problems.

When satisfactory equations can be formulated and sufficiently accurate parameters can be determined, – two matters of equal importance – a solution can be obtained, either by a digital computation or by an analytical method. An analytical method can often lead to a general result, whereas a computation can only apply to a particular numerical case. The analytical treatments are therefore still of considerable importance. The accuracy obtained depends on the error involved in the idealization of the machine on which the formulation is based and in the assumptions made in calculating the parameters.

Terminology and notation

The terminology and notation follows as far as possible the recommendations of recent international standards. The sign conventions are chosen, as befits a general theory, so that the equations apply equally to generators and motors. On the electrical side, they agree with the established conventions of circuit theory, as explained on p. 11.

The per-unit system is used extensively and extra care has been taken with the explanations, both in the main text and in Section 13.4.

The authors wish to acknowledge the encouragement and assistance from the Imperial College of Science and Technology and the University of Natal. Their thanks are also due to many colleagues and friends who have made useful suggestions, in particular to Brian Cory and Donald MacDonald for checking parts of the text and to June Harley, Allison Hudson, Wilhelmina van der Linden and Elizabeth Grant for typing the manuscript.

B. Adkins.

R. G. Harley.

The undersigned wishes to express his gratitude to the Central Electricity Generating Board, the South African Council for Scientific and Industrial Research and the Electricity Supply Commission of S.A., for sponsoring his research in the field of synchronous machine stability, particularly the last two organizations for their continued support of, and interest in, his work. Above all, he wishes to ackowledge the invaluable assistance of his wife June R. Harley, both tangible and intangible, given during the years of research and of writing this text.

R. G. Harley.

Chapter One
The Basis of the General Theory

1.1 The idealized machine

All types of rotating electromagnetic machine have many common features, as illustrated in Fig. 1.1. In the typical arrangement shown, there is an outer stationary member and an inner rotating member mounted in bearings fixed to the stationary member. The two elements carry cylindrical iron cores separated by an air-gap, and a main magnetic flux Φ passes across the air-gap from one core to the other in a closed magnetic circuit. Occasionally the inner member is fixed while the outer member rotates, or, in other special cases, both elements may rotate. The important feature of all rotating machines is that two cylindrical iron surfaces, a short distance apart, move relatively to one another. The cylindrical surface on either element may be virtually continuous, broken

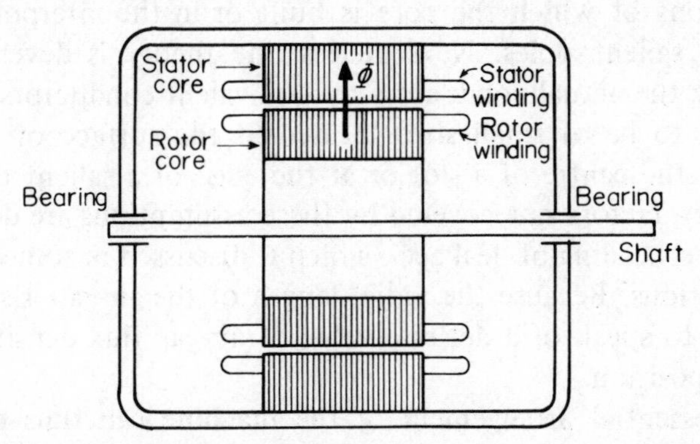

Fig. 1.1 Diagram showing the common features of all electrical machines.

only by small slot openings uniformly spaced round the circle, or it may be divided into an even number of salient poles with spaces between them.

Near the surface of each element, conductors carrying currents run parallel to the axis of the cylinders. The conductors are connected into coils by end connections outside the core, and the coils are connected to form the machine windings, which, on either element, consist of a relatively small number of circuits carrying independent currents. The operation of the machine depends primarily on the distribution of the currents round the core surfaces, and the analysis of the windings is concerned only with this distribution. The detailed arrangement of the end connections is of secondary importance. Thus the general arrangement is the same for all types of machine. The various types differ fundamentally only in the distribution of the conductors forming the windings and in whether the elements have continuous surfaces or salient poles. The operation of any given machine depends also on the nature of the voltages applied to its windings from external sources.

The narrow annular space between the two elements, known as the *air-gap*, is the critical region of the machine, and the theory is mainly concerned with the conditions in or near the air-gap. The magnetic flux is actually distributed in a complicated manner throughout the iron core, but, because of the high permeability of the iron, a reasonably simple and accurate theory can be derived by considering only the flux distribution in or near the air-gap. The conductors are actually located, not in the air-gap itself, but near to it. They are placed either in slots formed in the laminations of which the core is built or in the interpolar space between salient poles. Nevertheless, the theory is developed by replacing the actual conductors by equivalent conductors (usually assumed to be of small size) located at the surface of the core either at the centre of a slot or at the edge of a salient pole. The secondary factors not covered by these assumptions are dealt with under the heading of 'leakage', which is discussed in some detail in later sections. Because the radial length of the air-gap is small it is possible to speak of a definite value of air-gap flux density at any angular position.

The essential arrangement of the machine can thus be represented on a drawing of a section perpendicular to the axis of the cores. The two-dimensional drawing can then be further simplified to a *developed diagram* showing the air-gap circumference developed along a straight line. Fig. 1.2a shows, as an example, a sectional drawing of a six-pole machine having salient poles on the outer member and a core on the inner member which is continuous if the slots are ignored. Fig. 1.2b shows the corresponding developed diagram and indicates the fundamental sine wave of flux density. The developed diagram is exemplified in practise by the linear motor.

In the developed diagram the distribution of flux density and current repeats itself every two poles, whatever the actual number of poles may be. (Slight variations due to fractional-slot windings or staggered poles are ignored in working out the theory.) Hence any machine can be replaced by an equivalent two-pole machine

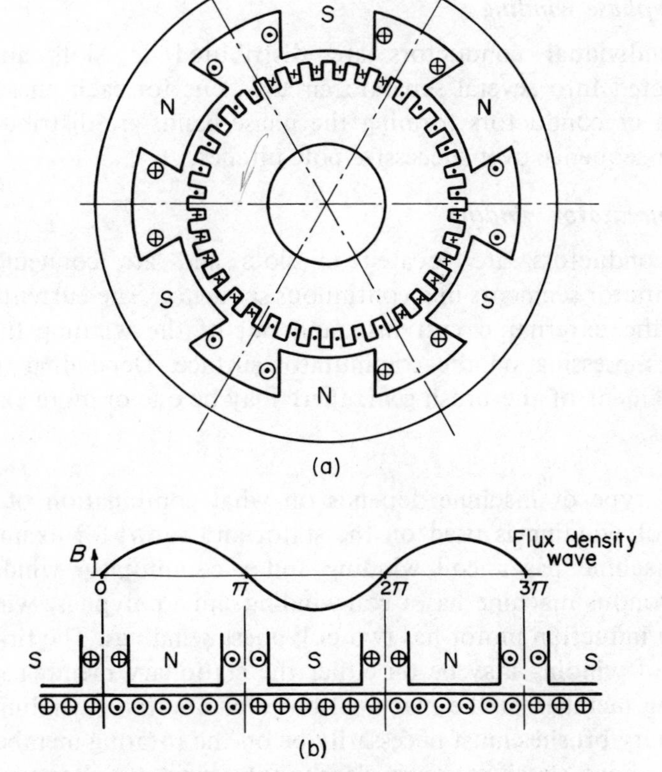

Fig. 1.2 A typical electrical machine.
(a) Sectional drawing. (b) Developed diagram.

and only such machines need be considered. In this book the theory is developed entirely in terms of two-pole machines. The number of poles must of course be introduced when calculating the constants of the machine, particularly when considering mechanical quantities such as torque and speed.

1.1.1 Types of winding

The windings of electrical machines are of three main types:

1. Coil winding

The winding consists of coils similarly placed on all poles, and connected together by series or parallel connection into a single circuit. Usually the coils are wound round salient poles as indicated in Fig. 1.2, but sometimes, as in a turbo-generator field winding, a coil winding may lie in slots.

2. Polyphase winding

The individual conductors are distributed in slots and are connected into several separate circuits, one for each phase. The groups of conductors forming the phase bands are distributed in regular sequence over successive pole pitches.

3. Commutator winding

The conductors are located in slots and are connected to commutator segments in a continuous sequence. The current flows from the external circuit into and out of the winding through brushes pressing on the commutator surface. Depending on the arrangement of the brush-gear, there may be one or more external circuits.

The type of machine depends on what combination of these types of winding is used on the stator and rotor; for example, a d.c. machine has a coil winding and a commutator winding, a synchronous machine has a coil winding and a polyphase winding, and an induction motor has two polyphase windings. The first two types of winding may be on either the stationary member or the rotating member of the machine, but a commutator winding with stationary brushes must necessarily be on the rotating member.

The above classification is clearly not a rigid one. For example, a single-phase a.c. winding can be treated alternatively as a special case of a polyphase winding or as a coil winding. A uniform cage winding, though not strictly a polyphase winding with independent circuits, can usually be replaced by one. An irregular winding cage, like the damper cage on a synchronous machine, on the other hand, must be treated as a number of separate coil windings.

1.1.2 The idealized two-pole machine

The practical machine is too complicated for an exact representation. Hence for any type of machine the theory is developed for an *idealized two-pole machine*, which is approximately equivalent to the practical machine. Each winding of the practical machine, or each part of a winding forming a circuit, is represented in the idealized machine by a single coil. For the purpose of analysis it is immaterial which of the two elements of the machine rotates and which is stationary, since its operation depends only on the relative motion between them. In a practical machine with salient poles, for example, either element may carry the salient poles. For a machine with a commutator, however, the commutator winding is always on the rotor, and the element carrying it is always indicated on the diagram of the idealized machine as the inner rotating member.

As an example, two alternative arrangements of an idealized three-phase salient-pole synchronous machine are shown in Fig. 1.3. A practical machine of this type usually has its three-phase armature winding on the stator and the field winding on the rotor as shown in Fig. 1.3a, but machines with the reverse arrangement as shown in Fig. 1.3b are often built. It is convenient when representing and analysing such a machine to use the form in Fig. 1.3b, but it is understood that the theory applies to either arrangement. Only one of the two salient poles is indicated.

The outer member in Fig. 1.3b carries a field coil F and damper coils KD and KQ. The axis of the pole round which the field coil F is wound is called the *direct axis* of the machine, while the axis 90° away from it is called the *quadrature axis.* At the instant considered, the axis of the coil representing the armature phase A makes an angle θ in a counter-clockwise direction with the direct axis. The instantaneous angular speed is

$$\omega = \frac{d\theta}{dt}$$

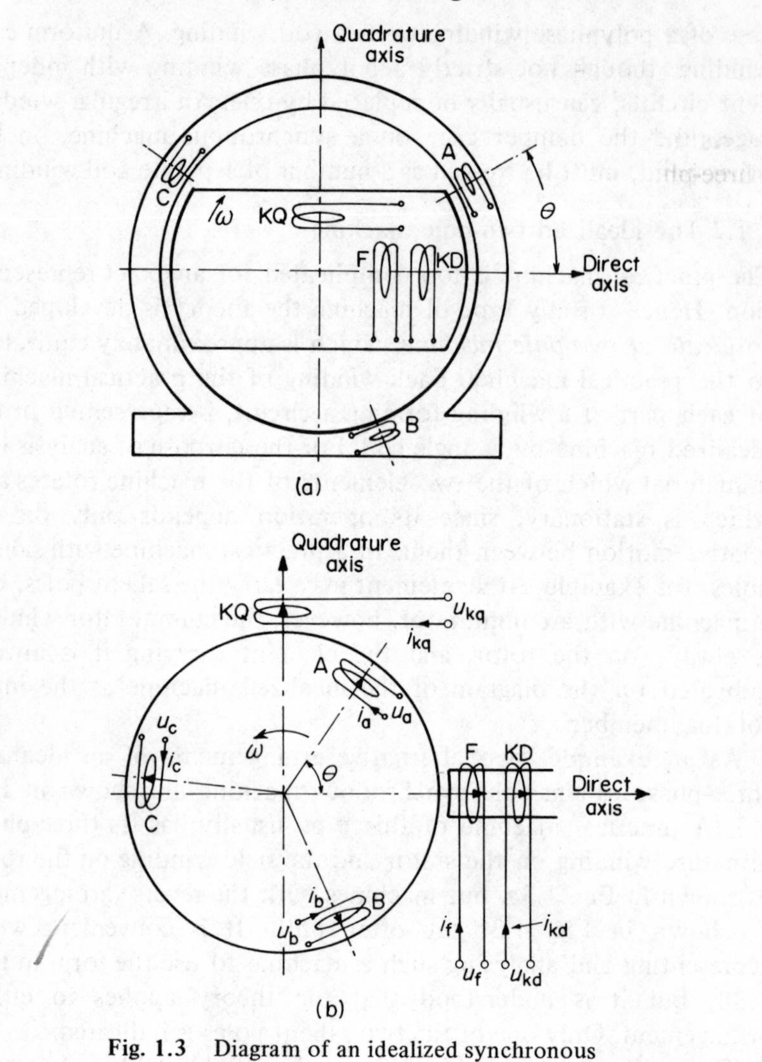

Fig. 1.3 Diagram of an idealized synchronous machine
(a) Armature winding on stator, and field winding and damper on rotor.
(b) Field winding and damper on stator, and armature winding on rotor.

The positive direction of the current in any coil is into the coil in the conductor nearer to the centre of the diagram, as indicated by an arrow in Fig. 1.3b. The positive direction of the flux linking a coil is radially outwards, as indicated by an arrow-head along the axis in the middle of the loop representing the coil. The loops of the coils are not intended to show the direction round the pole, but the convention is adopted that a current in a positive direction in a coil on any axis sets up a flux in the positive direction of that axis.

In referring to the diagrams of idealized machines, such as Fig. 1.3, the term *coil* is used to indicate a separate circuit in the machine which carries a single current. In the practical machine such a coil may consist of many turns, distributed over many poles and often in several slots on each pole. The term *winding*, on the other hand, may refer to a single coil – for example, the field coil F – or it may refer to several coils, as in the three-phase armature winding represented by the three coils A, B and C. Often it is necessary to make approximations. The cage damper winding of a practical synchronous machine consists of many circuits carrying different currents and would require a large number of coils for its exact representation. For many purposes it is sufficiently accurate to represent the damper winding by only two coils KD and KQ, which afford an example of the kind of approximation that is necessary in order to obtain a workable theory.

As a second example, Fig. 1.4 is the diagram of an idealized cross-field direct-current machine having salient poles on the stator and a commutator winding with main brushes on the direct

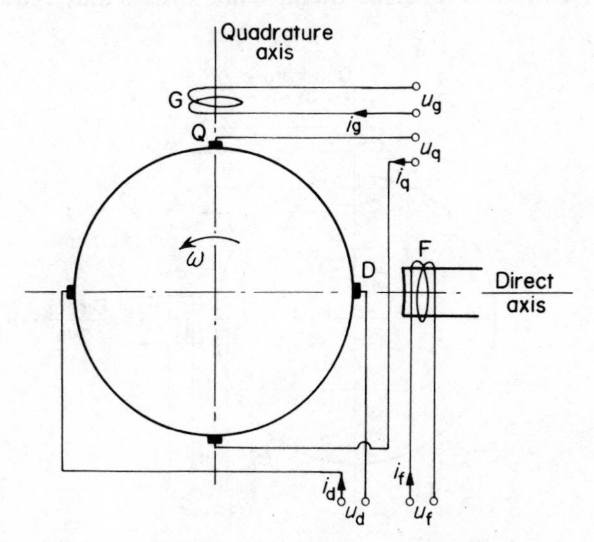

Fig. 1.4 Diagram of an idealized cross-field d.c. machine.

axis and cross brushes on the quadrature axis. In the diagrams of commutator machines the convention is adopted that a brush is shown at the position occupied by the armature conductor to which the segment under the brush is connected. The circuits through the brushes are labelled D and Q respectively. The machine represented also has a main field winding F and a quadrature field winding G acting along the direct and quadrature axes respectively. The machine may be represented alternatively by the four coils of Fig. 1.5, corresponding to the four separate circuits of Fig. 1.4.

The general theory given hereafter is developed to cover a wide range of machines in a unified manner. A very important part of this generalization applies to the two-axis theory in which, by means of an appropriate transformation, any machine can be represented by coils on the axes. However, other forms are possible, for example the phase equations of an a.c. machine or those of a commutator machine with brushes displaced from the axis. Kron called the two-axis idealized machine, from which many others can be derived, a *primitive* machine. Fig. 1.5 shows the diagram of such a primitive machine with one coil on each axis on each element, viz., F and G on the stationary member and D and Q on the rotating member. Some machines may require fewer than four coils to represent them, while others may require more.

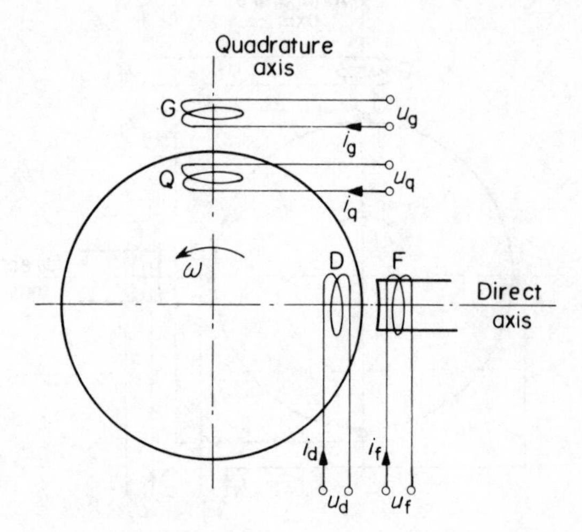

Fig. 1.5 Diagram of a primitive machine with four coils.

Any machine, however, can be shown to be equivalent to a primitive machine with an appropriate number of coils on each fixed axis. If the coils of the practical machine are permanently located on the axes, they correspond exactly to those of the primitive machine, but, if they are not, it is necessary to make a conversion from the variables of the practical machine to the equivalent axis variables of the corresponding primitive machine, or vice versa. Any particular system of description of a machine is known as a *reference frame* and a conversion from one reference frame to another is known as a *transformation.*

In order that the effects of rotation in a practical machine may be representable in an equivalent primitive machine, it is necessary, as shown later, that the coils D and Q representing the winding on the moving element of the primitive machine should possess special properties. The properties are the same as those possessed by a commutator winding, in which the current passes between a pair of brushes, viz.,

1. A current in the coil produces a field which is stationary in space.
2. Nevertheless a voltage can be induced in the coil by rotation of the moving element.

Such a coil, located on a moving element but with its axis stationary, and so possessing the above two properties, may be termed a *pseudo-stationary coil.*

The commutator machine of Fig. 1.4 may be either a d.c. machine or an a.c. commutator machine depending on the nature of the supply voltage. The actual circuits D and Q through the commutator winding not only form pseudo-stationary coils like the coils D and Q of the primitive machine of Fig. 1.5, but have the same axes as these coils. There is thus an exact correspondence between Figs. 1.4 and 1.5, and as a result the two-axis equations derived for the primitive machine of Fig. 1.5 apply directly to the commutator machine of Fig. 1.4. On the other hand, the three moving coils on the armature of the synchronous machine of Fig. 1.3b do not directly correspond to the coils D and Q of Fig. 1.5. In developing the two-axis theory of the synchronous machine, the three-phase coils A, B and C are replaced by equivalent axis coils D and Q, like those of Fig. 1.5, or in other words, the variables of the (a,b,c) reference frame are replaced mathematically by variables of the (d,q) reference frame. The transformation involved in the conversion, which depends on the fact that the same m.m.f. is set up in the machine by the currents of either reference frame, is explained and justified in Section 4.2.

In applying the two-axis theory to any type of rotating electrical machine the process may be summarized as follows. The diagram of the idealized two-pole machine is first set out using the smallest number of coils required to obtain a result of sufficient accuracy. The idealized two-pole machine is then related to a primitive machine with an appropriate number of coils, either directly or by means of suitable transformations. The theory is developed by deriving a set of voltage equations relating the voltages and currents of the primitive machine and, in addition, a torque equation relating the torque to the currents. The speed appears as a variable in the equations.

1.2 The two-winding transformer. Explanation of sign conventions and the per-unit system for electrical quantities

The process of idealization described in Section 1.1 reduces the machine to a set of mutually coupled coils, which, however, differ from normal mutually coupled coils because of the special pseudo-stationary property assigned to the coils on the rotating member. A machine, in fact, differs from a static transformer because of its rotation, but a consideration of the simpler apparatus which does not rotate is valuable as a starting point. In this section the two-winding transformer is used to explain several of the important concepts in the treatment of the rotating machine.

Consider the system of two mutually coupled coils illustrated in Fig. 1.6. Each coil is assumed to be *concentrated*, that is, its turns are wound closely together so that the same flux links every turn. A current in either coil flowing in the direction shown sets up a flux in the direction indicated by the heavy arrow.

1.2.1 Main flux and leakage flux

The flux in the transformer can be split up into three parts:

(a) *Main flux* Φ linking both coils.
(b) *Primary leakage flux* Φ_{11}, due to current i_1, and linking coil 1 but not coil 2.

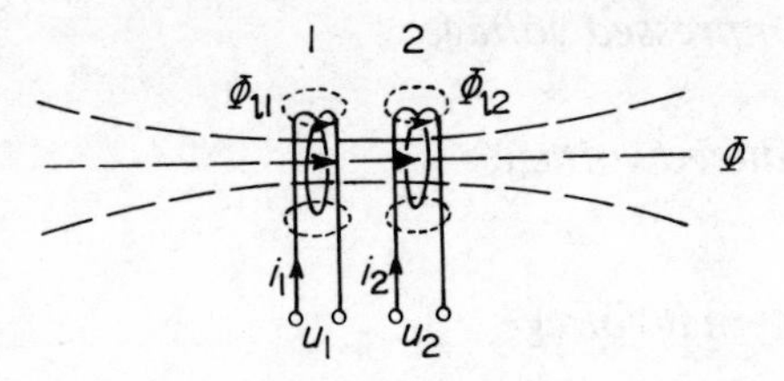

Fig. 1.6 Diagram of a transformer.

(c) *Secondary leakage flux* Φ_{12}, due to current i_2, and linking coil 2 but not coil 1.

This statement embodies the fundamental definition of the leakage flux of one coil in relation to another, viz., that it is the flux, due to the current in the one coil, which links that coil but not the other.

1.2.2 Sign conventions

The sign convention adopted for voltages and currents is as follows:

u represents the voltage impressed on the coil from an external source

i represents the current measured in the same direction as u

With the above convention, the instantaneous power ui flows into the circuit from outside, if both u and i are positive.

In a general theory covering both motors and generators it is important to use the same sign convention throughout. The convention adopted corresponds directly to *motor operation* and introduces negative quantities for generator operation. It agrees with the usual convention of circuit theory and has the advantage of introducing the minimum number of negative signs in the equations. It differs from the convention used by Park [3], who was concerned primarily with the synchronous generator, but agrees with that used by Kron [8], whose aim was to treat all the machines on a common basis.

To take a simple example, the equation for a circuit with resistance R and inductance L is:

$$u = Ri + L\frac{di}{dt}$$

where u is the *impressed voltage*

$-L\dfrac{di}{dt}$ is the *induced voltage*

$L\dfrac{di}{dt}$ is the *internal voltage*

1.2.3 The per-unit system

In developing the theories of electrical machines and power systems it is advantageous in many ways to express the quantities, not in actual units but in a *per-unit system.* To apply the per-unit system to the transformer of Fig. 1.6, the *base values of primary voltage and current* are chosen arbitrarily, generally as the nominal rated values. The *per-unit value* of voltage or current in the coil is then the actual value expressed as a fraction of the base value; for example, if the voltage is a half of rated voltage, $u = 0.5$ per-unit. The base value of time is one second.

The base values for the secondary quantities are related to the primary units. If the ratio of the secondary turns to primary turns is N, the *base value of secondary voltage* is N times that of the primary voltage, and the *base value of secondary current* is $1/N$ times that of the primary current. This definition is amplified in Section 13.4 where the base values of flux and inductance are also defined. It follows that in the per-unit system

$$\Phi = L_{12}(i_1 + i_2)$$
$$\Phi_{l1} = L_{l1}i_1$$
$$\Phi_{l2} = L_{l2}i_2$$

where L_{12} is the *mutual inductance* and L_{l1} and L_{l2} are the *leakage inductances.*

The induced voltage in coil 1 is:

$$-\frac{d}{dt}(\Phi + \Phi_{l1}).$$

The impressed voltage is in the opposite sense to the induced voltage. The value of the impressed voltage on coil 1 is:

$$u_1 = R_1 i_1 + (L_{12} + L_{l1})\frac{di_1}{dt} + L_{12}\frac{di_2}{dt} \tag{1.1}$$

There is a similar equation for the secondary coil. Using the symbol p for d/dt (see Section 13.3), the two equations of the transformer are:

$$\begin{bmatrix} u_1 \\ u_2 \end{bmatrix} = \begin{bmatrix} R_1 + (L_{12} + L_{l1})p & L_{12}p \\ L_{12}p & R_2 + (L_{12} + L_{l2})p \end{bmatrix} \begin{bmatrix} i_1 \\ i_2 \end{bmatrix} \quad (1.2)$$

1.2.4 Complete inductance and leakage inductance

The inductance of a coil is by definition given by the voltage induced by the total flux set up when the rate of change of current in the coil is unity. Applied to the primary coil this inductance would correspond to flux $(\Phi + \Phi_{l1})$ when the secondary winding carries no current ($i_2 = 0$). It is called the complete *self-inductance* and is represented by the symbol L_{11}. On the other hand, the leakage inductance L_{l1}, which in practical work is often called the inductance of the primary winding, is much smaller than the self-inductance, and corresponds only to the leakage flux Φ_{l1}. In terms of the self and mutual inductances, the equations for the system in Fig. 1.6 are

$$\begin{bmatrix} u_1 \\ u_2 \end{bmatrix} = \begin{bmatrix} R_1 + L_{11}p & L_{12}p \\ L_{12}p & R_2 + L_{22}p \end{bmatrix} \begin{bmatrix} i_1 \\ i_2 \end{bmatrix} \quad (1.3)$$

where

$$\left. \begin{aligned} L_{11} &= L_{12} + L_{l1} \\ L_{22} &= L_{12} + L_{l2} \end{aligned} \right\} \quad (1.4)$$

1.2.5 Advantages of the per-unit system

The per-unit system is of great benefit in making design calculations for machines, because it makes the comparison between different machines much easier. Corresponding quantities are of the same order of magnitude even for widely different designs.

In the formulation of the theory the per-unit system has the great merit that the numbers of turns do not enter into the equations. Moreover, the useful relation, stated in Eqns. 1.4, that the self-inductance of a coil is obtained by adding the mutual and leakage inductances, would not be true if the quantities were expressed in actual units.

1.3 Magneto-motive force and flux in the rotating machine

In any machine the currents in all the windings combine to produce the resultant flux. The action of the machine depends on the facts, firstly that the flux induces voltages in the windings, and secondly, that the flux interacts with the currents to produce torque.

1.3.1 Main air-gap flux

The flux spreads throughout the whole machine, but its effect depends primarily on the distribution of flux density round the air-gap; the attention is therefore focused on this region. At any instant the curve of flux density around the air-gap circumference may be of any form and is not necessarily sinusoidal. The *main flux* of an a.c. machine is defined to be that determined by the fundamental component of the curve of air-gap flux density, and the radial line where the fundamental density is a maximum is called the *axis of the flux.* The main flux is then completely defined by a magnitude and a direction.

1.3.2 Magneto-motive force

In order to calculate the flux due to a given system of currents it is first necessary to determine the magneto-motive force (m.m.f.) due to the currents. Fig. 1.7 is a developed diagram for a two-pole machine extending between angular positions 0 and 2π. The currents in the conductors of a coil are distributed in slots as indicated and form two bands, symmetrically distributed about

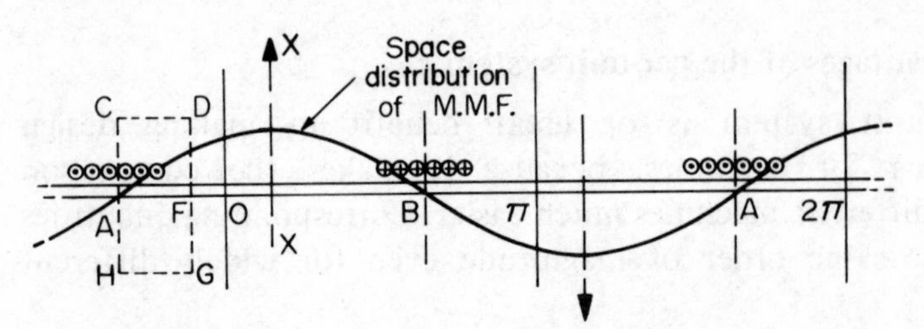

Fig. 1.7 Developed diagram showing the distribution of current and m.m.f.

the points A and B. The currents in the two bands flow in opposite directions. Since the current distribution is known, the m.m.f. round any closed path can be found; in particular, the m.m.f. round a path crossing the air-gap, such as ACDFGH. Because of the high permeability of the iron, all the m.m.f. round the closed path can be assumed to appear across the air-gap at the point F, if A is chosen as a point of zero air-gap flux density, which, for a symmetrical distribution of current, must occur at the centre of each band. Hence a curve of m.m.f. distribution round the air-gap can be drawn for any value of current flowing in the coil. Thus, although magneto-motive force is fundamentally a line integral round a closed path, a value can be associated with each point along the air-gap circumference, giving the space distributed *m.m.f. curve* of the machine. If the conductors are assumed to be located at points around the air-gap circumference, the m.m.f. curve is a stepped curve, but the fundamental component, which, by symmetry, is zero at the points A and B, can be drawn as in Fig. 1.7. The radial line at the point of maximum m.m.f. (*XX* in Fig. 1.7) is called the *axis of the m.m.f.*, and since this depends only on the conductor distribution, it is also the *axis of the coil.* The curve shows the instantaneous magnitude of the m.m.f., which depends on the instantaneous value of the current.

The m.m.f. curve determines the flux density curve. If the machine has a uniform air-gap, and if saturation is neglected, the flux density is everywhere proportional to the m.m.f. In a salient-pole machine, however, this is no longer true, and, in order to calculate the flux, it becomes necessary to resolve the m.m.f. wave into component waves along the direct and quadrature axes. On either axis a sinusoidal m.m.f. wave produces a flux density distribution, which can be determined by flux plotting or by other well-known methods. The flux density curve is not sinusoidal, but because of the symmetry of the pole about its axis, its fundamental component has the same axis as the m.m.f. producing it. Hence, if the harmonics are ignored, an m.m.f. wave on either the direct or quadrature axis produces a proportional sinusoidal flux density wave on the same axis, the factors of proportionality being different for the two axes. By this means the flux components on the two axes, due to any current, can be found and, if there is no saturation, the resultant flux, and hence the resultant flux density wave, is obtained by combining them.

1.4 Voltage and torque equations of the machine. The per-unit system for mechanical quantities

1.4.1 Steady and transient conditions

The equations of the primitive machine can be used to determine the performance of any of the different types of machine under any condition of operation, steady or transient. Generally the applied voltages are known and it is required to determine the currents, while the torque and speed may be either known or unknown quantities. It is not possible to obtain a general solution applicable to all conditions, and it is necessary to direct attention to particular problems.

The first important distinction is between transient and steady conditions. During a transient condition the voltages and currents, as well as torque and speed, are expressed as functions of time. The equations are differential equations and it is therefore necessary, for any particular problem, to know the initial conditions, or other boundary conditions, in order to obtain the solution. Steady conditions are of two types: d.c. conditions, when the quantities do not vary with time, and a.c. conditions, when the quantities vary sinusoidally with time. For steady conditions the general differential equations can be converted into algebraic equations, containing real or complex variables. In Section 5.1 the different types of problem encountered in practice are classified under seven headings (see Table 5.1) and the methods available for their solution are discussed.

The conventional theories of electrical machines apply mainly to steady conditions, which are easier to deal with than the more general transient conditions. The steady-state theories are usually developed for particular machines, in terms of phasor diagrams, equivalent circuits, and other devices, while transient conditions are considered quite independently. The *general theory* developed in this book embraces all these different conditions and shows the relation between the transient and steady conditions, as well as between the different types of machine.

1.4.2 The general voltage equations

The primitive machine of Fig. 1.5 has on each axis a pair of coils similar to the two coils of the transformer of Fig. 1.6. Since a stationary coil on one axis is mutually non-inductive with a stationary coil on the other axis, there would, if the machine were at rest, be no voltages induced in any coil on one axis due to currents in coils on the other axis. Hence the equations of each pair separately would be similar to those of the transformer. When the machine rotates, however, there are additional terms in the equations because, as a result of the rotation, voltages are induced in the pseudo-stationary coils D and Q by fluxes set up by currents on the other axis (see p. 9).

For the commutator machine of Fig. 1.4, or the alternative representation by four coils as shown in Fig. 1.5, the equations relating the voltages and currents in the four circuits are derived in Section 2.1 and stated in Eqn. (1.5). Parameters in Eqn. (1.5) may be either in actual units or in per-unit.

$$\begin{bmatrix} u_f \\ u_d \\ u_q \\ u_g \end{bmatrix} = \begin{bmatrix} R_f + L_{ff}p & L_{df}p & & \\ L_{df}p & R_d + L_d p & L_{rq}\omega & L_{rg}\omega \\ -L_{rf}\omega & -L_{rd}\omega & R_q + L_q p & L_{qg}p \\ & & L_{qg}p & R_g + L_{gg}p \end{bmatrix} \begin{bmatrix} i_f \\ i_d \\ i_q \\ i_g \end{bmatrix} \tag{1.5}$$

The constants in the equations are resistances R, self-inductances L_{ff}, L_d, L_q, L_{gg} and mutual inductances L_{df}, L_{qg} with suffixes indicating the coils to which they refer. L_{df} and L_{qg} are analogous to the mutual inductance L_{12} in the transformer Eqns. (1.3). L_{rd}, L_{rf}, L_{rq}, L_{rg} are additional constants of a similar nature to inductances, determining the voltage induced in an armature coil on one axis due to rotation in the flux produced by a current in a coil on the other axis.

In setting down the equations for each circuit in Fig. 1.5, the terms for all the internal voltages are added to the resistance drop and equated to the impressed voltage. The use of the matrix notation makes it easy to compare the different coefficients and shows at a glance which currents have zero coefficients.

The four equations given by Eqns. (1.5) refer to a machine represented by the four coils of Fig. 1.5. In general, a machine represented in the diagram by n coils has n voltage equations, relating n voltages to n currents. The impedance matrix of such a machine has n^2 compartments.

In Eqns. (1.5) some terms contain the derivative operator p, and represent voltages due to changing currents in coils on the same axis as the one being considered. They are called *transformer voltages*, and are present even when the machine is stationary. Other terms, containing the speed ω, represent voltages induced by rotation in the flux set up by the current in a coil on the other axis. For example, $(-L_{rf}\omega i_f)$ is the voltage induced between the quadrature-axis brushes by a current in the direct-axis field winding, and is the normal induced voltage in a simple d.c. machine consisting of circuits Q and F only. Such voltages are called *rotational voltages*. When the coils carry steady d.c. currents, there are no transformer voltages but the rotational voltages are still present.

With the conventions adopted on p. 6 the signs of the rotational voltage terms are as shown in Eqns. (1.5). The sign of any rotational voltage can be determined by considering the torque produced by the interaction of the fields set up by the currents in the windings. A positive current i_f sets up a flux passing outwards along the direct axis and hence a south pole on the surface of the stator pole, while a positive current i_q sets up a north pole on the surface of the armature. The attraction of these unlike poles leads to a torque on the armature in the negative direction, corresponding to generator action if ω is positive. Hence, with the convention of p. 6 the associated component of electrical input power must be negative, so that the rotational voltage term $(-L_{rf}\omega i_f)$ has the negative sign. The term $(-L_{rd}\omega i_d)$ also has a negative sign, because positive currents i_d and i_f produce fields in the same direction. A similar argument shows that the rotational voltages in the direct-axis armature circuit due to quadrature-axis currents have a positive sign.

In Section 2.1 the derivation of the equations of a d.c. machine is discussed more fully. In Chapter 4 it is shown that the equations of an a.c. machine can be expressed in the same form as those of a d.c. machine, if the variables are the fictitious axis voltages and currents determined by Park's transformation, instead of the actual phase quantities. Thus the equations for the primitive machine apply equally to a.c. or d.c. machines.

1.4.3 The general torque equation

At any instant the torque developed by the machine depends on the currents flowing in the windings. An equation for the torque can be deduced from the voltage equations by considering the instantaneous power. In the following derivation of Eqn. (1.10) all the quantities are in per-unit.

The torque developed by the interaction between the flux and the currents is called the *electrical torque*, and differs from the externally applied torque, if the speed varies, because of the inertia of the machine.

Let M_e be the instantaneous electrical torque, defined as positive when mechanical power is passing into the machine from outside at a positive speed;

M_t be the instantaneous applied torque (any friction or mechanical damping torque is assumed to be included with M_t);

J be the moment of inertia.

Then:

$$M_t = M_e + J\frac{d\omega}{dt} \tag{1.6}$$

The convention adopted following Kron [8] is intended to bring out the analogy between the electrical and mechanical quantities. Voltage is applied to the terminals and torque is applied to the shaft. Hence, apart from the effect of losses, positive voltage and current mean motoring action, and positive torque and speed mean generating action.

1.4.4 Mechanical units in the per-unit system

The total power P supplied to a machine electrically is the sum of the powers supplied to the individual circuits. (The capital letter P is used for the instantaneous power in order to avoid confusion with the operator p.) In the per-unit system it is desirable that the power input should have a value equal or close to unity when the voltage and currents in the main circuits have unit values. For machines having several main circuits it is therefore necessary to introduce a factor k_p in the formula for power and to choose the value of k_p according to the particular type of machine. Then

$$P = k_p \Sigma ui. \tag{1.7}$$

The *base value of power* is defined in Appendix 13.4 for a single main circuit as the power corresponding to base voltage and base current. In a system with more than one main circuit, for example, for a three-phase machine, base power is that supplied when base current flows at base voltage and at unity power factor in all three phases. Hence for a three-phase machine the value of k_p is $\frac{1}{3}$. For a cross-field generator, in which only one of the armature circuits is a main circuit, the base power is the power given by base current at base voltage in this circuit, and hence $k_p = 1$. In general k_p is the reciprocal of the number of main circuits.

It may be noted that, with the above definition, base power in an a.c. machine is not the actual rated power, but the power corresponding to rated kVA. Thus for an a.c. machine having a rating of 1000 kVA at 0.8 power factor, base power equals 1000 kW. The rated power is 800 kW, or 0.8 per-unit.

The *base value of speed* adopted here is the radian per second. This accords with the radian as the *base value of angle* and the second as the *base value of time*. Any machine has a *nominal speed*, denoted by ω_0 which can be used as a basis of reference. In a d.c. machine the nominal speed is the normal rated speed of the equivalent two-pole machine in 'electrical' radians per second. In an a.c. machine the nominal speed is the synchronous speed of the equivalent two-pole machine, and thus equals 2π times the supply frequency.

The *base value of torque* is defined as the torque which produces base power at the nominal speed ω_0 and hence in the per-unit system torque is numerically equal to power when speed ω equals ω_0. In practical work the moment of inertia of the machine is expressed by the *inertia constant*, denoted by H, which is a normalized quantity, but differs from the per-unit inertia J. H is defined by the following formula:

$$H = \frac{\text{Stored energy at synchronous speed in kW-sec}}{\text{Rated kVA}}$$

The inertia constant, which is usually calculated from the moment of inertia and the speed in normal units, has the dimensions of time and its value is given in seconds. The *base value of energy* is given by base power acting for one second, and hence H is numerically equal to the per-unit stored energy. If the machine were accelerated at constant acceleration from rest to the nominal speed ω_0 in one second, the torque would be, from Eqn. (1.6), equal to $J\omega_0$. It is shown on p. 22 that, with the above definition of base torque, the per-unit power at speed ω is numerically $P = (\omega/\omega_0)M$. Since the mean speed during the run up is $\omega_0/2$ and the time taken is one second, the per-unit stored energy is

$$H = \tfrac{1}{2}J\omega_0$$

or

$$J = \frac{2H}{\omega_0}$$

and Eqn. (1.6) becomes

$$M_t = M_e + \frac{2H}{\omega_0} p\omega \tag{1.8}$$

The per-unit system has the advantage, as mentioned on p. 13, that the quantities for different machines are of the same order of magnitude. The inertia constant H, which is used mainly for synchronous machines, has a value from two to six seconds for a wide range of designs of different sizes and speeds. The per-unit system, as applied to the mechanical quantities, has the disadvantage that the dimensional consistency of the equations is lost.

1.4.5 Calculation of electrical torque

Considering only the rotor coil D of Fig. 1.5, the power P_d supplied to this coil is, from Eqn. (1.5)

$$P_d = k_p u_d i_d$$

$$= k_p\left(R_d i_d^2 + L_d i_d \frac{di_d}{dt} + L_{df} i_d \frac{di_f}{dt} + L_{rq}\omega i_d i_q + L_{rg}\omega i_d i_g\right)$$

In the above expression for P_d, the first term is the ohmic loss, and the second and third terms give the rate of change of stored magnetic energy in the machine. Only the fourth and fifth terms contribute to the output power corresponding to the electrical torque. It therefore follows that the power P_e, corresponding to the torque developed by the interaction between flux and current, is determined by adding together the terms obtained by multiplying all the rotational voltages by the corresponding currents.

Hence for the machine represented in Fig. 1.5:

$$P_e = k_p \omega (L_{rq} i_d i_q + L_{rg} i_d i_g - L_{rf} i_q i_f - L_{rd} i_d i_q). \tag{1.9}$$

Using the definition of electrical torque given on p. 19, and the definition of base torque on p. 20,

$$P_e = -\frac{\omega}{\omega_0} M_e$$

The negative sign appears because P_e is derived from the electrical power passing into the terminals, whereas M_e is defined as a torque applied to the shaft. The following expression is therefore obtained for the electrical torque:

$$M_e = \omega_0 k_p [L_{rf} i_q i_f - L_{rg} i_d i_g + (L_{rd} - L_{rq}) i_d i_q]. \tag{1.10}$$

Eqn. (1.10) applies particularly to a machine represented by the four coils in Fig. 1.5, for which the equations are those stated in Eqns. (1.5). In the more general case the same method can be used, although there may be more or fewer terms depending on the number of rotational voltage terms in the equations, and the expression for M_e can be written down for any machine in the same way. If then the value of M_e is substituted in Eqn. (1.8), an equation is obtained giving M_t in terms of the currents and the speed.

1.5 The fundamental assumptions. Saturation, harmonics, leakage

The assumptions made in Section 1.3 are based on those given by Park [3]. The principal assumptions are that there is no saturation and that space harmonics in the flux wave may be neglected. Before proceeding with the theory, it is worth while to examine what is involved.

1.5.1 Saturation and other non-linear effects

In the idealized machine it is assumed that all flux densities are proportional to the currents producing them; that is, that there are no saturation or other non-linear effects. Now in practice saturation is an important factor, and much ingenuity has been used in devising methods of taking it into account. These methods occupy a large part in any textbook on electrical machine design; for example, the treatment of armature reaction in a d.c. machine, or the use of 'Potier reactance' in a synchronous machine. Such methods do not, however, introduce the non-linear property into the basic theory. They are directed mainly to the determination of appropriate values of constants to suit the particular problem, the constants being defined in relation to a linear theory.

The fundamental theory must of necessity be based on linear relationships, because a theory which allowed directly for non-linear effects introduces much complication, and it is important that the practical adjustments should not be allowed to obscure the fundamental background of the theoretical analysis. The present book is primarily concerned with the development of the linear theory. Methods of allowing for the effect of saturation are discussed in more detail in Chapter 10.

In applying the theory to practical problems the effect of saturation must always be borne in mind, since the accuracy of the final result depends on the validity of the methods used. Because of the uncertainty introduced by the presence of saturation it is more than ever important to verify the results obtained, whenever possible, by tests on actual machines, so as to develop methods of determining the appropriate constants [7]. For the study of systems in which the machines are very large, experiments with micro-machine models are extremely valuable because of the difficulty of carrying out tests on very large equipment.

With the assumption of linearity the *principle of superposition* can be used. The current in any coil sets up a magneto-motive force and hence a component of flux, which may induce a voltage in the coil itself or in any other coil. By the principle of superposition all the component voltages in any coil can be added together to obtain the resultant voltage. The voltage equations for all the coils, together with a torque equation constitute the *general equations of the machine*, which determine completely its operation when the applied voltages and torque are known.

1.5.2 Harmonics

The neglect of space harmonics in the general theory, although convenient for giving a simple explanation, is actually a good deal more drastic than is necessary, because the effect of many harmonics can be included by modifying the values of the main and leakage inductances. When the machine is at rest, all the space harmonics in the flux wave can be allowed for in this way, since it is then in effect a static transformer. When the machine rotates some of the voltages due to harmonics would require additional, and usually more complicated, terms in the equations, and it is these harmonics that are neglected.

Consider first a machine operating with alternating current under steady conditions. The phasor diagram for any particular circuit can only include voltages of a single frequency. Any space harmonics which induce voltages of this frequency can therefore be allowed for in the theory, while those which induce voltages of a different frequency must be neglected. The harmonics of the second group produce parasitic effects such as noise, voltage ripples, or parasitic torques, and it is important to make them as small as possible. A good deal of attention has been paid to this question, and a good design embodies many special features for the purpose of reducing undesirable harmonics. Such matters are outside the scope of this book, and it is assumed in the development of the two-axis theory that these harmonics can be neglected.

The general equations of a machine apply to all possible manners of variation of the instantaneous values of current and voltage, among which a sine wave alternation is only a special case. The flux harmonics, however, still fall into two groups, namely, those which induce voltages that can be included with the normal terms in the equation, and those which would require additional terms. It is only the harmonics of the second group that are neglected.

1.5.3 Leakage

The greater part of the leakage flux in a machine, defined in the same way as for the transformer on p. 10, arises because the conductors are located in slots instead of at the air-gap surface, and because the windings have end connections outside the core. Considering first a machine having only one winding on each element, the main leakage flux around the slots does not cross the air-gap and consequently only links the winding which produces it. There is also the so-called *zigzag leakage* flux, which crosses the air-gap but does not penetrate far enough into the core to link the winding on the other side. The inductance corresponding to these types of leakage can be calculated by well-known methods [30].

In addition, a winding produces harmonic fluxes which cross the air-gap and link the other winding, but are distinct from the main flux which has been defined as the fundamental part. In some instances the voltages induced by these harmonics can be usefully allowed for in the equations of the machine, and in others they cannot. For example, in an induction-motor running under steady conditions, the voltage induced in one winding by a harmonic flux wave due to the other winding cannot be included in the equations. This is because the voltage has a different frequency from the voltage induced by the corresponding, fundamental flux wave. The voltage induced by the harmonic flux wave, can only be allowed for when the effect of the harmonic is equivalent to an increase of leakage flux linked with the winding. The equivalent increase of leakage flux is known as *belt leakage.*

In a commutator machine, on the other hand, the commutator winding produces harmonic flux waves which are similar to those set up by a stator winding. Such harmonics normally induce voltages in the stator of such a nature that they can be included in the equations, and consequently their effect is equivalent to an increase of the main flux.

It is therefore not true that all harmonics are neglected in the two-axis theory, provided that the correct values of inductances are chosen. Those harmonics that are neglected can, however, cause considerable inaccuracy if the machine is not well designed or if the harmonics are deliberately increased in order to obtain special properties (Chapter 12). It is important in a normal machine to reduce the harmonics by correctly choosing the number and size of the slots, the pitch of the coils, and by other well-known precautions, so that all harmonic winding factors can be assumed to be zero.

1.5.4 Distributed windings

In the transformer discussed in Section 1.2 the coils are assumed to be concentrated and, as a result, there is a definite line of demarcation between main flux and leakage flux. If, on the other hand, the coils consist of turns distributed in space in different positions, there is no clear distinction between the two parts of the flux, because it is not possible to say what flux links both coils. One way of expressing this variability is to take the primary leakage reactance X_a to be to some extent arbitrary. It is generally accepted that the standard test methods for synchronous machines cannot determine X_a separately. The matter is important when it is required to determine the fundamental mutual and leakage reactances of a synchronous machine from the measured values of the transient and subtransient reactances and time constants. (See Section 4.7). It is necessary to select a value of X_a, usually a design value calculated from the dimensions, after which all the other parameters have definite values.

It is sometimes suggested that the Potier reactance, which is an effective leakage reactance allowing empirically for saturation under steady conditions, should be used as the value for X_a in transient calculations. The two conditions are however quite different and the Potier reactance is usually a good deal too high for the purpose.

In the transformer, the effective number of turns is taken to be the obvious number surrounding the core and there is no difficulty. In a rotating machine, where the armature winding is always distributed, the ambiguity is removed by defining the main flux as the fundamental sine wave component of the flux density curve. The effective number of turns is equal to the actual number, multiplied by a *winding factor* based on the assumed sinusoidal flux density wave.

If there are three coils, the use of leakage becomes more complicated. The straightforward method is to write the equations with three self and three mutual reactances, showing that six independent parameters are necessary to determine the system completely. Since however in practical machines the six parameters only differ from each other by small amounts, it becomes necessary for the sake of accuracy to express them in terms of a single mutual reactance and, at most, five much smaller quantities which depend on the differences, and are of the nature of leakages. The matter is discussed further in Section 4.7.

1.6 Calculation and measurement of parameters

The matrix equations of a machine, for example Eqns. (1.5), provide a definition of each parameter, since each element of the impedance matrix determines the voltage that would be obtained if only one current were present. The methods of calculating and measuring each parameter are based on this separation of the components.

Design calculations of inductances are made by assuming current to flow in one winding only and determining the m.m.f., the flux, and finally the induced voltage, in all the windings [22]. The flux pattern is determined with simplifying assumptions and the final formula almost always requires some empirical adjustment based on practical experience. Like the transformer inductances in Eqn. (1.3), each self-inductance in Eqn. (1.5) can be divided into a mutual component and a leakage component, which are calculated separately. Leakage calculations are usually made by considering the flux passing round a slot, assuming the iron to have infinite permeability, and empirical corrections are made to compensate the error in this assumption and to allow for the leakage of the end winding. Separate calculation of the leakage is necessary because many performance characteristics depend more on the leakage inductances than on the mutual inductances. Details of the calculations for particular machines are outside the scope of this book.

Measurement techniques [39] depend as far as possible on the same principle, but there are many circuits, like those of a cage winding, where it is not possible to pass a separate current from outside. Moreover, even when it is possible, as in a slip-ring induction motor, direct measurement of the self and mutual inductances would not determine the leakage inductances with sufficient accuracy, for the reasons stated above. Consequently, tests with two or more windings in action, for example a short-circuit test, become necessary. Nevertheless the principle behind all test methods is to choose the simplest possible practical condition, involving the smallest number of currents. Again the details of test methods are outside the present scope. A good fundamental discussion about the measurement of parameters is given in reference [34].

Chapter Two
The Primitive Machine

2.1 The equations of the cross-field commutator machine

The analysis of a cross-field commutator machine, of which the conventional d.c. machine is a special case, is included here because of the clear physical separation of the two axes. The variables in the primitive machine are identical with those in the practical machine and no transformation is needed. The theory applies directly to a.c. commutator machines, with or without a transformation. The two-axis theory of a.c. machines without commutators in later chapters will be followed more easily if the reader has a sound understanding of the material in the present chapter.

The voltage equations of the cross-field d.c. machine represented in Fig. 1.4 were stated without full proof in Eqns. (1.5). The result can be justified by considering each term separately as being the voltage which would arise in the circuit concerned if there were current in only one of the four circuits. The terms for resistance drops or voltages induced in stator windings need no further explanation, but the armature voltages, already discussed on p. 18, are considered in more detail in the following pages. The sign convention used has been explained in Section 1.2.

2.1.1 Transformer voltages in the armature

The transformer voltage induced between the brushes on either axis is proportional to the rate of change of the flux on the axis.

The transformer voltage in the direct-axis armature circuit due to the main air-gap flux Φ_{md} may be written

$$u_{\mathrm{dt}} = p\psi_{\mathrm{md}} \tag{2.1}$$

where ψ_{md} is the *flux linkage*, corresponding to Φ_{md}, with the armature circuit. ψ_{md} and Φ_{md} both depend on the curve of flux density in the air-gap. They are proportional to each other and the relation between them depends on the distribution both of the flux density and of the armature winding. The transformer voltages can thus be expressed by using inductance coefficients as in Eqns. (1.5). If the distribution of the flux density is assumed to be sinusoidal, certain relations hold between the inductance coefficients which are of importance in connection with the two-axis theory (see Eqns. (2.8)).

The transformer voltage between the direct-axis brushes due to a direct-axis flux can be found by calculating the voltages induced in individual turns and integrating over half the circumference. In Fig. 2.1a the dots and crosses on the conductors indicate the directions in which the armature conductors are traversed as the circuit is followed through in the positive direction between the direct-axis brushes. A positive current in this circuit would set up a flux along the positive direct axis in accordance with the convention stated on p. 6.

If the direct-axis flux density curve is sinusoidal, with maximum value B_{m}, then

$$B = B_{\mathrm{m}} \cos \theta$$

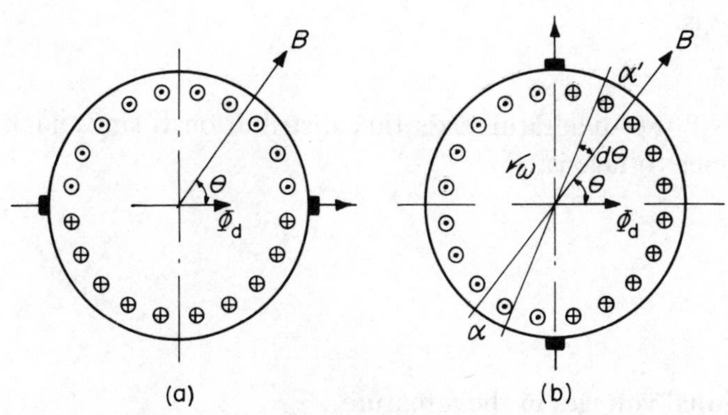

Fig. 2.1 Directions of the conductors in the armature circuits.
(a) Direct axis. (b) Quadrature axis.

and

$$\Phi_{\mathrm{m\,d}} = \int_{-\pi/2}^{\pi/2} Blrd\theta = 2B_{\mathrm{m}} lr$$

where

B = air-gap flux density at the point θ,

r = radius of the armature,

l = length of the armature core.

The flux linking a turn whose sides are located at the points θ and $(\pi + \theta)$ is:

$$\int_{(\pi+\theta)}^{\theta} B_{\mathrm{m}} \cos\theta \,.\, lrd\theta = 2B_{\mathrm{m}} lr \sin\theta = \Phi_{\mathrm{m\,d}} \sin\theta$$

The induced transformer voltage in the turn is:

$$-\frac{\mathrm{d}}{\mathrm{d}t}(\Phi_{\mathrm{m\,d}} \sin\theta)$$

The internal transformer voltage between the direct-axis brushes, which is the negative of the induced voltage, is given by:

$$u_{\mathrm{d\,t}} = \int_{0}^{\pi} \frac{\mathrm{d}}{\mathrm{d}t}(\Phi_{\mathrm{m\,d}} \sin\theta)\frac{z\mathrm{d}\theta}{2}$$

$$= \frac{\mathrm{d}}{\mathrm{d}t}(\Phi_{\mathrm{m\,d}} z)$$

where z = number of conductors per radian.

Hence, from Eqn. (2.1), the flux linkage for a sinusoidal flux density wave is:

$$\psi_{\mathrm{m\,d}} = \Phi_{\mathrm{m\,d}} z.$$

Similarly, if the quadrature-axis flux distribution is sinusoidal, the transformer voltage is:

$$u_{\mathrm{q\,t}} = p\psi_{\mathrm{m\,q}} \qquad (2.2)$$

where

$$\psi_{\mathrm{m\,q}} = \Phi_{\mathrm{m\,q}} z$$

2.1.2 Rotational voltages in the armature

The *rotational voltage* induced between the quadrature-axis brushes due to the armature rotation through a direct-axis flux, can be found by calculating the voltages in the individual conductors and integrating over half the circumference. In Fig. 2.1b the dots and crosses on the conductors indicate the directions in which the armature conductors are traversed as the circuit is followed through in the positive direction between the quadrature-axis brushes.

The voltage in the positive direction in the conductors of coil $\alpha\alpha'$ occupying a small part of the periphery subtending an angle $d\theta$ is, by the flux-cutting rule:

$$B \cdot l \cdot r\omega \cdot z \cdot d\theta .$$

The direct-axis flux is given by:

$$\Phi_{md} = \int_{-\pi/2}^{\pi/2} Blrd\theta .$$

The internal rotational voltage between the quadrature-axis brushes, which is the negative of the induced voltage, is therefore:

$$\begin{aligned} u_{qr} &= - \int_{-\pi/2}^{\pi/2} Blr\omega z d\theta \\ &= -\omega z \Phi_{md} \\ &= -\omega \psi_{md} \end{aligned} \qquad (2.3)$$

Similarly the rotational voltage between the direct-axis brushes due to a quadrature-axis flux can be found in the same way. The expression is of the same form as Eqn. (2.3) but now has a positive sign (see p. 18):

$$u_{dr} = \omega \psi_{mq} . \qquad (2.4)$$

The result expressed by Eqns. (2.3) and (2.4) does not depend at all on the wave-form of the flux density curve. The field form curve of a d.c. machine does in practice depart considerably from a sinusoidal shape, but the rotational voltage depends only on the area of the curve, which is proportional to the total flux per pole. ψ_{mq} appears in both Eqns. (2.2) and (2.4) only because a sinusoidal flux density was assumed in deriving Eqn. (2.2). The same condition applies for ψ_{md}.

2.1.3 Armature voltage equations

Eqns. (2.1) to (2.4) give the transformer and rotational voltages due to sinusoidally distributed air-gap fluxes Φ_{md} and Φ_{mq} on the two axes, each flux being measured by the corresponding flux linkage ψ_{md} or ψ_{mq}. In order to obtain the armature voltage equations the effect of the armature leakage must also be included. To do this it is assumed that the armature leakage flux acts in the same way as a sinusoidally distributed air-gap flux so far as both rotational and transformer voltages in the armature winding are concerned. Then, if ψ_d and ψ_q are the total direct-axis and quadrature-axis flux linkages with the armature, including leakage, the armature voltage equations are:

$$\left.\begin{aligned} u_d &= p\psi_d + \omega\psi_q + R_a i_d \\ u_q &= -\omega\psi_d + p\psi_q + R_a i_q \end{aligned}\right\} \tag{2.5}$$

where R_a is the armature resistance between a pair of brushes and is assumed to have the same value for either axis.

The flux linkages are related to the currents by

$$\left.\begin{aligned} \psi_d &= L_d i_d + L_{df} i_f \\ \psi_q &= L_q i_q + L_{qg} i_g \end{aligned}\right\} \tag{2.6}$$

where L_d, L_q are the self-inductances of the armature circuits; L_{df}, L_{qg} are the mutual inductances between each armature circuit and the field winding on the same axis.

The voltage equations of the machine are thus

$$\begin{array}{|c|}\hline u_f \\ \hline u_d \\ \hline u_q \\ \hline u_g \\ \hline \end{array} = \begin{array}{|c|c|c|c|}\hline R_f + L_{ff}p & L_{df}p & & \\ \hline L_{df}p & R_d + L_d p & L_q\omega & L_{qg}\omega \\ \hline -L_{df}\omega & -L_d\omega & R_q + L_q p & L_{qg}p \\ \hline & & L_{qg}p & R_g + L_{gg}p \\ \hline \end{array} \begin{array}{|c|}\hline i_f \\ \hline i_d \\ \hline i_q \\ \hline i_g \\ \hline \end{array} \tag{2.7}$$

They are of the same form as Eqns. (1.5) but some of the coefficients in Eqns. (1.5) are now equal in pairs because of the sinusoidal flux density distribution. Hence if

$$\left.\begin{aligned} L_{rd} &= L_d, & L_{rf} &= L_{df} \\ L_{rq} &= L_q, & L_{rg} &= L_{qg} \end{aligned}\right\} \tag{2.8}$$

are substituted into Eqns. (1.5) they become Eqns. (2.7).

The torque (Eqn. (1.10)) can now be expressed in terms of the flux linkages.

$$M_e = k_p \omega_0 (\psi_q i_d - \psi_d i_q) \qquad (2.9)$$

2.2 Application to a simple d.c. machine

As a practical example of the primitive machine, consider the simplest possible d.c. machine illustrated by Fig. 2.2. The machine has only two windings. The stator has a field winding on the vertical axis (which is now the pole axis of the d.c. machine) and the rotor has a commutator winding with brushes located so that an armature current sets up a field on the horizontal axis. The vertical axis is chosen as the pole axis in order to avoid having a negative sign in the voltage equations. The diagram represents a separately excited d.c. machine with its brushes in the neutral

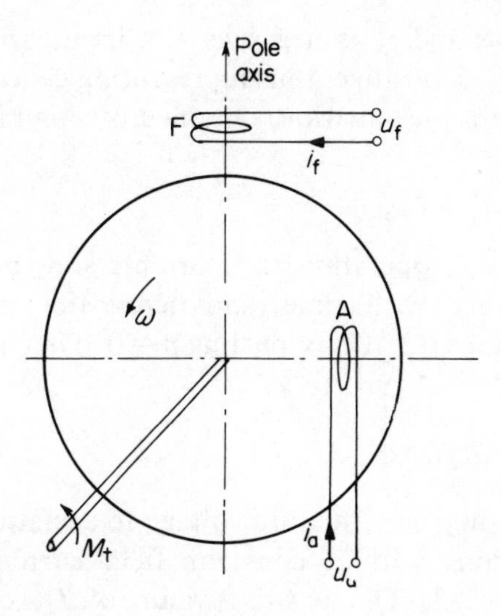

Fig. 2.2 Diagram of a simple d.c. machine:

position. As in Fig. 1.5, the armature circuit through the brushes can be replaced by an axis coil on the direct axis. The coil A then has the pseudo-stationary property discussed on p. 9.

The coils F and A of Fig. 2.2 correspond to the coils G and D in the primitive machine of Fig. 1.5 and the equations can be found from Eqns. (1.5) by omitting the first and third rows and columns of the impedance matrix.

$$\begin{bmatrix} u_f \\ u_a \end{bmatrix} = \begin{bmatrix} R_f + L_{ff}p & \\ L_{af}\omega & R_a + L_a p \end{bmatrix} \begin{bmatrix} i_f \\ i_a \end{bmatrix} \qquad (2.10)$$

These two equations do not depend on any assumption that the flux wave is sinusoidal, since both the rotational voltage $L_{af}\omega i_f$ in the armature winding and the voltage $L_{ff}pi_f$ in the field winding depend on the total flux regardless of its distribution. Eqns. (2.10) hold whether the quantities are expressed in actual units or in per-units.

The electrical torque is obtained by substituting $i_d = i_a$, $i_q = 0$, $\psi_q = L_{af}i_f$, and $k_p = 1$, in Eqn. (2.9).

$$M_e = -\omega_0 L_{af} i_f i_a \qquad (2.11)$$

If i_f is positive and i_a is negative, the input armature power is negative and M_e is positive, thus representing generator action.

The overall torque equation, obtained by combining Eqns. (1.6) and (2.11), is

$$M_t = -\omega_0 L_{af} i_f i_a + Jp\omega \qquad (2.12)$$

For steady d.c. operation the currents and speed, denoted by ω_s, do not vary with time, and hence the equations can be derived from Eqns. (2.10) by putting $p = 0$. They are

$$\left.\begin{aligned} U_f &= R_f I_f \\ U_a &= L_{af}\omega_s I_f + R_a I_a \end{aligned}\right\} \qquad (2.13)$$

The curve relating the armature voltage to armature current under steady conditions with a constant field current and speed, is shown in Fig. 2.3. The positive value of I_a, corresponding to motoring action, is plotted to the left in Fig. 2.3 in order to show the characteristic for a separately excited d.c. generator in the form usual in textbooks. For the idealized machine the curve is a straight line.

The effects of brush shift, series windings, interpole and compensating windings can be allowed for by addition of suitable coils to the primitive machine of Fig. 2.2. However the treatment of such phenomena is not within the scope of this book. See [20].

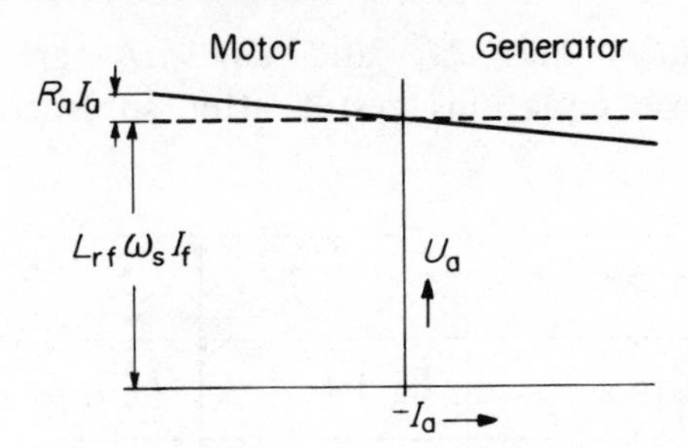

Fig. 2.3 Characteristic of a d.c. machine.

2.3 Equations for small changes and small oscillations

The last two types of problem covered by Table 5.1, p. 99, arise when a machine operating under steady conditions is subjected to a small disturbance. If the changes in the variables are small, so that their squares or products can be neglected, the differential equations relating the changes are linear even when the general equations are non-linear. Such equations can be used for studying steady-state stability, which depends on the effect of making a small change relative to a steady condition, or for calculating the magnitude of small oscillations which may be superimposed on a condition of steady operation.

In the present section the method of analysing small changes or small oscillations is explained in detail for the simple d.c. machine represented in Fig. 2.2 for which the equations are Eqns. (2.10) and (2.12). When ω is variable, these equations are non-linear because they contain the products ωi_f and $i_f i_a$. However, they can be linearized by neglecting the products or squares of the small changes. The technique of linearization is explained here in relation to the simple d.c. machine. The method is applied in Section 7.6 to the more complicated synchronous machine.

Assume that the field voltage changes from a steady value u_{f0} to a slightly different value $u_{f0} + \Delta u_f$, and that all the other variables change similarly. Eqns. (2.10) and (2.12) become

$$\left.\begin{aligned} u_{f0} + \Delta u_f &= (R_f + L_{ff}p)(i_{f0} + \Delta i_f) \\ u_{a0} + \Delta u_a &= L_{af}(\omega_s + \Delta\omega)(i_{f0} + \Delta i_f) + (R_a + L_a p)(i_{a0} + \Delta i_a) \\ M_{t0} + \Delta M_t &= -\omega_0 L_{af}(i_{f0} + \Delta i_f)(i_{a0} + \Delta i_a) + Jp(\omega_s + \Delta\omega) \end{aligned}\right\} \tag{2.14}$$

If the products $\Delta\omega \cdot \Delta i_f$ and $\Delta i_f \cdot \Delta i_a$ are neglected, the following linear equations result after subtracting the original equations

$$\begin{bmatrix} \Delta u_f \\ \Delta u_a \\ \Delta M_t \end{bmatrix} = \begin{bmatrix} R_f + L_{ff}p & & \\ L_{af}\omega_s & R_a + L_a p & L_{af}i_{f0} \\ -\omega_0 L_{af} i_{a0} & -\omega_0 L_{af} i_{f0} & Jp \end{bmatrix} \begin{bmatrix} \Delta i_f \\ \Delta i_a \\ \Delta\omega \end{bmatrix} \quad (2.15)$$

The above process is similar to that of partial differentiation and the third equation of Eqns. (2.15) could equally well have been found as follows:

$$\begin{aligned} \Delta M_t &= -\omega_0 L_{af}\Delta(i_f i_a) + J\Delta(p\omega) \\ &= -\omega_0 L_{af}(i_f \Delta i_a + i_a \Delta i_f) + Jp(\Delta\omega) \end{aligned}$$

after which i_f and i_a are replaced by i_{f0} and i_{a0} to indicate that these are the original steady values. The differentiation is carried out with respect to the dependent variables and not with respect to t; hence the operator p is not affected by it. The point is brought out clearly by the long method, based on first principles, used above to derive Eqns. (2.14) and (2.15).

The differential Eqns. (2.15) can be used for studying stability by imagining a sudden small change to be made in any of the applied quantities u_a, u_f or M_t and finding the resulting change in i_a, i_f or ω. Whichever of these variables is determined, the operational expression for the variable has a denominator of the form f(p) given by the determinant of the impedance matrix in Eqn. (2.15). Hence

$$\mathrm{f}(p) = (R_f + L_{ff}p)[Jp(R_a + L_a p) + \omega_0 L_{af}^2 i_{f0}^2]$$

For stability the equation $\mathrm{f}(p) = 0$ must have no roots with positive real parts, and it can be investigated by any of the well-known stability criteria.

2.3.1 Phasor equations for small oscillations

If the small changes considered above vary sinusoidally with time at a frequency $m/2\pi$, Eqns. (2.15) can be converted into phasor equations by replacing p by jm and replacing the instantaneous values Δu etc., by the complex quantities ΔU etc. The phasor equations, which can be written down straight away from the general equations, are:

$$\begin{bmatrix} \Delta U_f \\ \Delta U_a \\ \Delta M_t \end{bmatrix} = \begin{bmatrix} R_f + jmL_{ff} & & \\ L_{af}\omega_s & R_a + jmL_a & L_{af}i_{f0} \\ -\omega_0 L_{af} i_{a0} & -\omega_0 L_{af} i_{f0} & jmJ \end{bmatrix} \begin{bmatrix} \Delta I_f \\ \Delta I_a \\ \Delta\Omega \end{bmatrix} \quad (2.16)$$

A d.c. generator driven by a diesel engine provides a simple example of a problem that can be solved by this method. Assume that the torque pulsates at a frequency $m/2\pi$ and that the armature and field voltages are constant. The torque pulsation, which is superimposed on the normal steady torque, is represented by the complex number ΔM_t. The voltage pulsations are zero, and hence $\Delta U_f = \Delta U_a = 0$. The first equation (2.16) shows that ΔI_f is also zero (that is, that the flux is constant). The equations reduce to:

$$\begin{bmatrix} 0 \\ \Delta M_t \end{bmatrix} = \begin{bmatrix} R_a + jmL_a & L_{af}i_{f0} \\ -\omega_0 L_{af} i_{f0} & jmJ \end{bmatrix} \begin{bmatrix} \Delta I_a \\ \Delta\Omega \end{bmatrix} \quad (2.17)$$

$\Delta\Omega$ and ΔI_a determine the magnitude and phase of the pulsations of speed and armature current.

2.4 Sudden short-circuit of a d.c. generator

The problem of calculating the transient current in a d.c. generator which is suddenly short-circuited is used here to explain the method, which is applied in Chapter 8 to the more complicated synchronous machine. It is an example of the third type of problem in Table 5.1, p. 99. It is assumed that the speed remains constant during the short-circuit [13].

When the short-circuit is applied the armature current rises rapidly to a high peak and then dies away slowly to the steady short-circuit value (see Fig. 2.5). The field current also rises to a peak and then dies away. Because of the heavy currents involved it is essential to allow for the effect of armature reaction.

In a machine with no saturation, the m.m.f. due to the armature current has no effect on the main flux in the poles. There is, however, a good deal of saturation in most d.c. machines, particularly when heavy currents flow, and as a result the main flux is reduced because of the cross magnetisation. The effect is not by any means linearly proportional to the current and cannot strictly be taken into account in a linear theory. Fig. 2.4 shows the steady load characteristic of a d.c. generator, in which the voltage falls to zero at five times full load. As an approximation, the curve can be replaced by a straight line, corresponding to a greatly increased armature resistance R, determined by its slope. During a transient change, the changing flux produces an additional term in the field voltage equation proportional to di_a/dt. Hence with these rather drastic assumptions the equations are

$$\left.\begin{aligned} u_f &= (R_f + L_{ff}p)i_f + L_{fs}pi_a \\ u_a &= L_{af}\omega i_f + (R + L_a p)i_a \end{aligned}\right\} \tag{2.18}$$

where L_{fs} is the mutual inductance representing the transient effect of armature reaction.

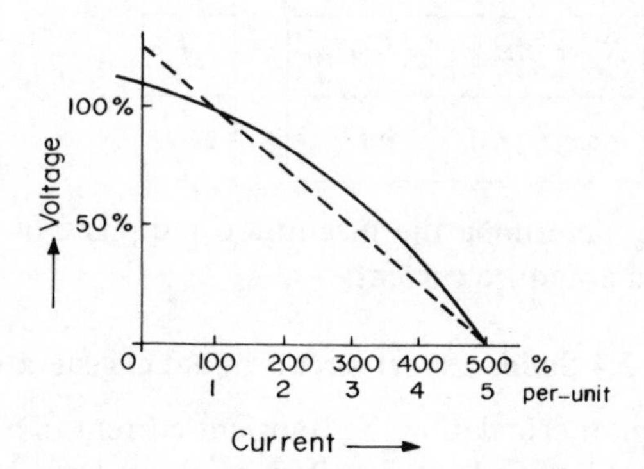

Fig. 2.4 Load characteristic of a d.c. machine.

Consider a generator excited at a constant field voltage and operating unloaded before the short-circuit. The effect of the short-circuit is to reduce the armature voltage u_a suddenly to zero, while leaving the field voltage unchanged. The constant value of the armature voltage before the short-circuit is denoted by U. It is required to find how the armature current varies with time after the short-circuit.

The problem is best handled by means of the principle of superposition. The voltages and currents after the short-circuit are each equal to the sum of the original value and the change resulting from the short-circuit. Thus:

$$i_f = i_{f0} + i_f'$$
$$i_a = i_{a0} + i_a'$$

where i_{f0}, i_{a0} are the original values, ($i_{a0} = 0$, in the example). i_f', i_a' are the superimposed currents and are functions of time.

In this and later examples, the word *original* is used for the values existing before the change takes place, while the word *initial* is used for the value immediately after the change.

The field voltage does not change, and hence the superimposed voltage u_f' is zero. On the other hand, the armature voltage is U before the short-circuit and changes abruptly to zero at the instant of short-circuit. Because the equations are linear, they are satisfied by the superimposed quantities, and hence become:

$$\left.\begin{aligned} 0 &= (R_f + L_{ff}p)i_f' + L_{fs}p i_a' \\ -U &= k i_f' + (R + L_a p) i_a' \end{aligned}\right\} \qquad (2.19)$$

Since the initial values of the superimposed currents are zero, Eqn. (2.19) becomes, after transformation into the Laplace domain, (see Appendix 13.3)

$$\begin{bmatrix} 0 \\ -\dfrac{U}{p} \end{bmatrix} = \begin{bmatrix} R_f(1 + T_f p) & L_{fs}p \\ k & R(1 + T_a p) \end{bmatrix} \begin{bmatrix} \bar{i}_f' \\ \bar{i}_a' \end{bmatrix} \qquad (2.20)$$

where

$k = L_{af}\omega_s$

$T_f = L_{ff}/R_f$ = (field time constant)

$T_a = L_a/R$ = (effective armature time constant)

In the present problem, the actual short-circuit current i_a is the same numerically as i_a' because the original current i_{a0} is zero, and it is negative because the machine operates as a generator. The positive value, from Eqns. (2.20) is

$$\bar{i}_s = -\bar{i}_a = \frac{(1 + T_f p)}{(1 + T_a p)(1 + T_f p) - \dfrac{L_{fs} k p}{R_f R}} \cdot \frac{U}{Rp} \tag{2.21}$$

$$= \frac{(1 + T_f p)}{(1 + T_a' p)(1 + T_f' p)p} \cdot \frac{U}{R} \tag{2.22}$$

Using the standard form of Eqn. (13.11), with the approximation that $T_f, T_f' > T_a$, the solution is

$$i_s = U[1/R + (1/R' - 1/R)\,\epsilon^{-t/T_f'} - (1/R') \cdot \epsilon^{-t/T_a'}] \tag{2.23}$$

R' is called the *transient resistance.* The time constants and R' are given by

$$T_a' = \frac{L_a}{R'}, \qquad T_f' = T_f\left[1 - \frac{kL_{fs}}{RL_{ff}}\right], \qquad R' = \frac{T_f'}{T_f} \cdot R.$$

Fig. 2.5 (full line) shows a typical curve of short-circuit current plotted from Eqn. (2.23). The upper curve (dotted in the initial portion) represents the sum of the first and second terms, and is what would be obtained if the armature inductance were zero. The current given by this curve rises instantly to U/R' and then dies away to the steady current U/R with time constant T_f'. Due to the armature inductance, the actual current rises, not instantaneously, but at a rapid rate depending on T_a'. The peak value is, however, still given approximately by U/R' but is slightly less. The initial rate of rise is U/L_a.

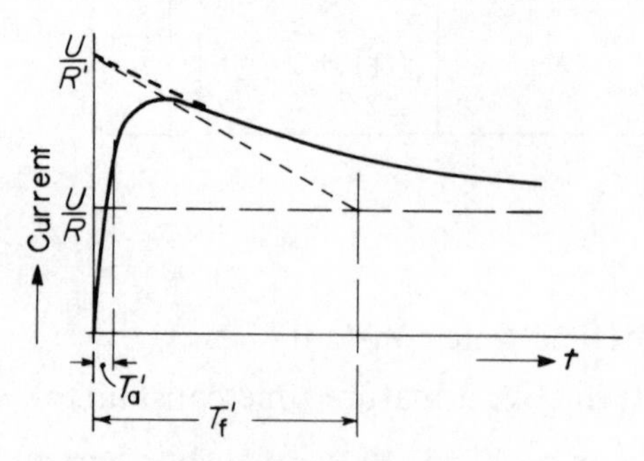

Fig. 2.5 Sudden short-circuit of a d.c. generator.

Although the assumptions are such that the results obtained are only approximate, the above theory gives a satisfactory qualitative explanation of the behaviour of a d.c. generator after a short-circuit. It is virtually impossible to calculate the parameters of a given machine from first principles, but a technique for calculating them can be built up from the analysis of short-circuit and other tests.

Example. Tests were taken on a 130-volt, 300-amp separately excited d.c. generator, and analysed so as to obtain the parameters.

Load test $U_a = 130, I_a = -300, I_f = 3.5, U_f = 130$

Steady short-circuit test $U_a = 0, I_a = -1500, I_f = 3.5$

Using these figures in Eqns. (2.18), and putting $p = 0$, gives

$$130 = 3.5R_f$$

$$130 = 3.5L_{af}\omega_s - 300R$$

$$0 = 3.5L_{af}\omega_s - 1500R$$

whence

$$R_f = 37.2, R = 0.108, L_{af}\omega_s = 46.5 = k$$

Sudden short-circuit test. The oscillogram of armature current agreed approximately with the following curve:

$$i_s = 1205 + 2530\epsilon^{-t/0.089} - 3260\epsilon^{-t/0.0023}$$

giving, by comparison with Eqn. (2.25)

$$\frac{U}{R} = 1205, \qquad \frac{U}{R'} = 3260, \qquad T_f' = 0.0023$$

The following parameters are deduced from the tests:

$$R = 0.108, \ R' = 0.04, \ T_f = 0.24$$

$$L_a = 0.092 \times 10^{-3}, \ L_{ff} = 8.9, \ L_{fs} = 0.013, \ T_a = 8.51 \times 10^{-6}$$

Chapter Three
The Steady-State Phasor Diagrams of A.C. Machines

3.1 Representation of sinusoidal m.m.f. and flux waves by space phasors

Before deriving the general equations of a.c. machines it is worth while to consider briefly the special case of steady operation, which can be analysed independently by means of phasor equations and diagrams. The method used in this chapter, which is essentially the same for the different types of a.c. machine, follows the same lines as the later development of the general theory, but is a good deal simpler. The present chapter forms a link between the usual textbook treatment and the general treatment which is the main theme of the book.

The theory starts with the currents flowing in the various windings of the machine and determines first the m.m.f. and then the flux produced by the currents. From the flux, the voltages induced in the windings are calculated and hence the phasor diagrams or equations are obtained. For an a.c. machine operating under steady conditions with balanced polyphase voltages and currents, the fundamental flux density wave rotates round the air-gap at constant speed relative to the a.c. winding. The steady-state phasor diagrams are based on a consideration of the rotating waves of m.m.f. and flux density produced by the polyphase currents.

The phasors in the voltage and current diagrams are called *time phasors*, because the quantity represented varies sinusoidally with time. *Space phasors* on the other hand represent quantities which vary sinusoidally in space; for example, the fundamental m.m.f. around the air-gap circumference. At any instant a sinusoidal wave of m.m.f. can be represented by a space phasor with its axis at the point of maximum m.m.f.; two or more such waves can be added by phasor addition.

A space phasor of m.m.f. is defined by a complex number $\boldsymbol{F}$ which is related to the space distribution of the m.m.f. in the same way that a time phasor of current, defined by the complex number $\boldsymbol{I}$, is related to the time variation of the current (see Appendix 13.1). Fig. 3.1 is a developed diagram showing the variation of the m.m.f. with the angle θ. If F_m is the maximum m.m.f., the value at the point θ is:

$$F = F_m \cos(\theta + \phi) = \mathrm{Re}[\mathbf{F}\epsilon^{j\theta}] \qquad (3.1)$$

where

$$\boldsymbol{F} = F_m\, \epsilon^{j\theta}$$

Thus F is a complex number analogous to that used for a time phasor except that, following the usual practice, the magnitude of the space phasor is taken to be equal to the maximum value, rather than the r.m.s. value, so that the factor $\sqrt{2}$ does not appear in Eqn. (3.1). For the machines dealt with in the present chapter the flux and m.m.f. waves are represented by space phasors, while the voltages and currents are represented by time phasors.

The voltage induced in a coil by a rotating or stationary flux wave is proportional to the flux and to the relative speed between the flux and the coil. The time phase of the voltage depends on the space position of the flux wave relative to the coil. Hence the time phasor representing the induced voltage is directly related to the space phasor representing the flux.

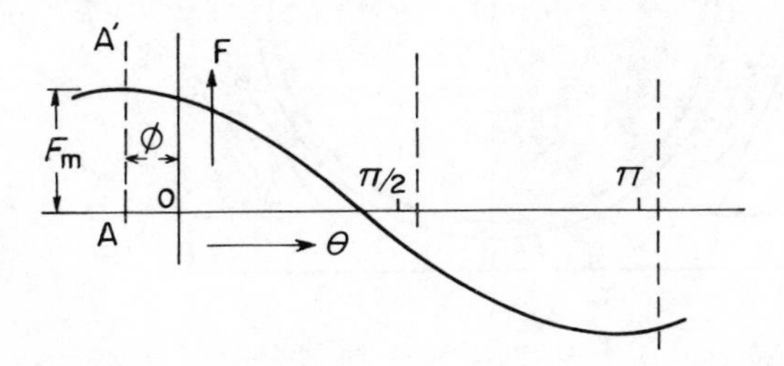

Fig. 3.1 Curve of air-gap m.m.f.

When a machine has a uniform air-gap, the air-gap flux density is everywhere proportional to the m.m.f. and the induced voltage can be obtained directly from the m.m.f. In a salient-pole machine, on the other hand, the constant proportionality no longer holds. When, however, the m.m.f. wave rotates at the same speed as the poles, it can be resolved into two components, one along each axis. The corresponding components of the flux wave can then be found as explained on p. 15. Hence the salient-pole synchronous machine requires a *two-axis theory.* The usual theory of the uniform air-gap synchronous machine and of the induction motor does not depend on resolving the flux into the axis components and is thus a true *rotating field theory.* However, they can also be studied as a special case of the two-axis theory.

3.2 The induction motor

Fig. 3.2 is a diagram of an idealized two-pole induction motor having phase coils A_1, B_1, C_1 on the stator and A_2, B_2, C_2 on the rotor (for clearness only coil A_2 is shown in the diagram).

The induction motor has no obvious axis of symmetry since both members are cylindrical. In Fig. 3.2 the position of the quadrature axis has been chosen arbitrarily to coincide with the axis of the stationary coil A_1. The phase sequence is A_1 $-C_1$ $-B_1$.

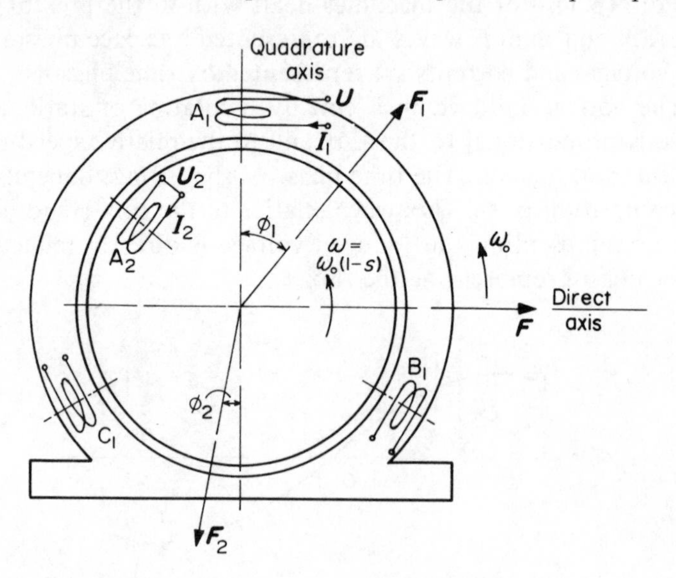

Fig. 3.2 Diagram of an induction motor.

Coil A_2 rotates with the rotor at the constant speed $\omega_0(1-s)$, where ω_0 is the synchronous speed and s is the slip. During steady operation the m.m.f. and flux waves due to the balanced polyphase currents in each winding rotate relative to the stator at synchronous speed ω_0 and combine to form the resultant air-gap m.m.f. and flux waves. The instant at which the resultant m.m.f. wave, represented by the space phasor $\boldsymbol{F}$, is on the direct axis (drawn horizontally), is indicated on Fig. 3.2. At the same instant, the component m.m.f.s due to the primary and secondary currents respectively are represented by the space phasors $\boldsymbol{F}_1$ and $\boldsymbol{F}_2$ at angles $(-\phi_1)$ and $(180° - \phi_2)$ to the vertical. Since the three phasors move round together at speed ω_0, they can be related by a stationary phasor diagram (Fig. 3.3a).

If core losses are neglected, the resultant flux wave due to $\boldsymbol{F}$ also has its axis on the direct axis, and consequently the internal voltage, opposing that induced in the primary phase A_1, has its maximum value at the instant considered. The internal voltage is represented by the phasor $\boldsymbol{U}_i$. If the primary windings carried polyphase currents such that the current in phase A_1 was in phase with $\boldsymbol{U}_i$, they would set up an m.m.f. on the quadrature axis at this instant. Since, however, the phasor $\boldsymbol{F}_1$ is displaced by ϕ_1 from the vertical axis, the corresponding primary current $\boldsymbol{I}_1$ actually lags behind $\boldsymbol{U}_i$ by the angle ϕ_1 (see Fig. 3.3b). In the primary voltage phasor diagram of Fig. 3.3b, which applies to phase A_1,

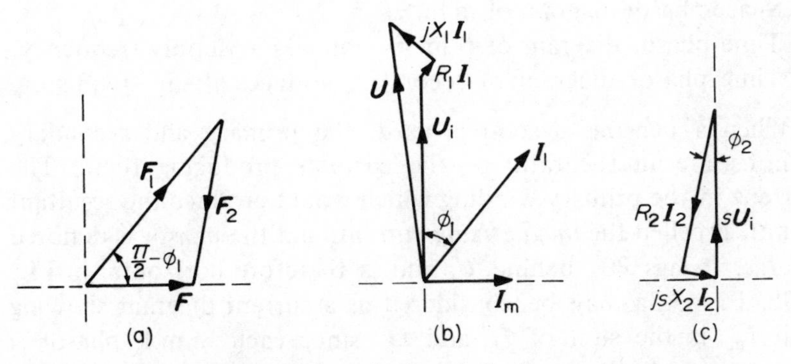

Fig. 3.3 Phasor diagrams of a polyphase induction motor.
(a) Space diagram of m.m.f.s.
(b) Time diagram of primary voltages.
(c) Time diagram of secondary voltages.

the terminal impressed voltage $\boldsymbol{U}$ is the sum of the internal voltage $\boldsymbol{U}_{\mathrm{i}}$, the resistance drop $R_1\boldsymbol{I}_1$ and the leakage reactance drop $jX_1\boldsymbol{I}_1$.

The secondary winding is not located in any definite position relative to the primary winding and in general has a different number of turns per phase. If a per-unit system, defined for the stationary induction motor in the same way as for the transformer on p. 12 is used, the voltage induced in phase A_2 by the resultant flux has the same magnitude as $\boldsymbol{U}_{\mathrm{i}}$ if the rotor is at rest. With a slip s the flux induces a slip-frequency voltage, while the phase depends on the rotor position. In Fig. 3.3c the phasor $s\boldsymbol{U}_{\mathrm{i}}$, like $\boldsymbol{U}_{\mathrm{i}}$ in Fig. 3.3b, represents the voltage opposing the induced voltage. If the secondary winding carried polyphase currents such that the current in phase A_2 was in phase with the secondary induced voltage, the secondary currents would set up an m.m.f. on the quadrature axis at the instant when the flux is along the direct axis. This relation holds whatever the actual position of the secondary winding may be at that instant. Since, for the operating condition indicated in Fig. 3.2, $\boldsymbol{F}_2$ is displaced from the quadrature axis by the angle $(\pi - \phi_2)$, it follows that the secondary current phasor $\boldsymbol{I}_2$ is displaced by $(\pi - \phi_2)$ from $s\boldsymbol{U}_{\mathrm{i}}$, as shown in the time phasor diagram of Fig. 3.3c. In the secondary voltage phasor diagram (Fig. 3.3c), the sum of $s\boldsymbol{U}_{\mathrm{i}}$, the resistance drop $R_2\boldsymbol{I}_2$, and the leakage reactance drop $jsX_2\boldsymbol{I}_2$, is zero, since the secondary winding is short-circuited.

The theory of the induction motor is thus based on three separate phasor diagrams.

1. Space phasor diagram of m.m.f.s.
2. Time phasor diagram of primary voltages at supply frequency.
3. Time phasor diagram of secondary voltages at slip frequency.

When a per-unit system is used, the primary and secondary m.m.f.s are in the ratio of the currents producing them. The current in the primary winding which would produce the resultant m.m.f. is called the *magnetizing current*, and the phasor is denoted by $\boldsymbol{I}_{\mathrm{m}}$. It lags 90° behind $\boldsymbol{U}_{\mathrm{i}}$ and is therefore horizontal in Fig. 3.3b. Fig. 3.3a may be considered as a current diagram showing that $\boldsymbol{I}_{\mathrm{m}}$ is the sum of $\boldsymbol{I}_1$ and $\boldsymbol{I}_2$, since each m.m.f. phasor is proportional to the corresponding current phasor.

The phasor equations can be written down from the diagrams

(Figs. 3.3a–c). The quotient of the magnitudes of U_i and I_m is a constant called the *magnetizing reactance* X_m. The equations are:

$$\left.\begin{aligned} U &= U_i + (R_1 + jX_1)I_1 \\ 0 &= sU_i + (R_2 + jsX_2)I_2 \\ I_m &= I_1 + I_2 \\ U_i &= jX_m I_m \end{aligned}\right\} \qquad (3.2)$$

Elimination of U_i and I_m from these equations gives the two voltage equations of the induction motor:

$$\begin{bmatrix} U \\ 0 \end{bmatrix} = \begin{bmatrix} R_1 + j(X_m + X_1) & jX_m \\ jsX_m & R_2 + js(X_m + X_2) \end{bmatrix} \begin{bmatrix} I_1 \\ I_2 \end{bmatrix} \qquad (3.3)$$

Eqns. (3.3) are similar to those of the transformer. If Eqns. (1.2) are converted into phasor equations, and new symbols are introduced for the reactances, they correspond exactly to Eqns. (3.3), except for the presence of the slip s. With the motor at rest ($s = 1$) the equations become identical with those of the transformer.

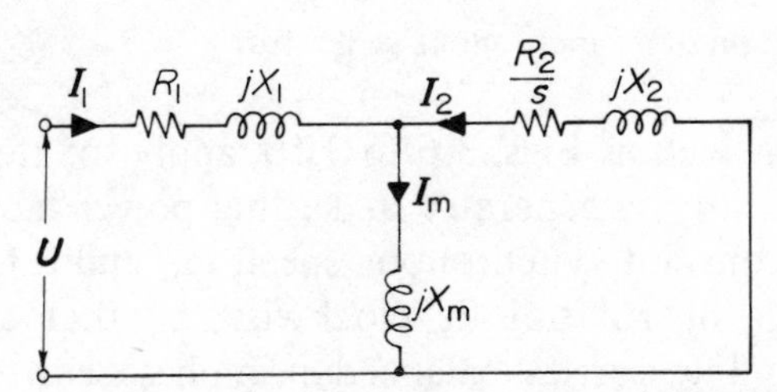

Fig. 3.4 Equivalent circuit of an induction motor.

Fig. 3.4 shows an *equivalent circuit* of the induction motor. An equivalent circuit of a machine may be defined as a system of static elements in which the currents and voltages satisfy the same equations as the machine. It is evident that Eqns. (3.3) apply to the network shown in Fig. 3.4. It may be noted that the direction of I_2 is opposite to that usually shown in textbooks.

3.3 The uniform air-gap synchronous machine

Fig. 3.5 is a diagram of a two-pole synchronous machine having a field winding F on the outer member and a three-phase winding on the inner rotating member. In a practical machine, particularly in a large generator, the field system is almost always on the rotor, but in the diagram of the idealized machine it is indicated on the stator and its axis is taken as the direct axis. The damper windings shown on Fig. 1.3 are omitted in Fig. 3.5 because they do not affect the operation under steady conditions.

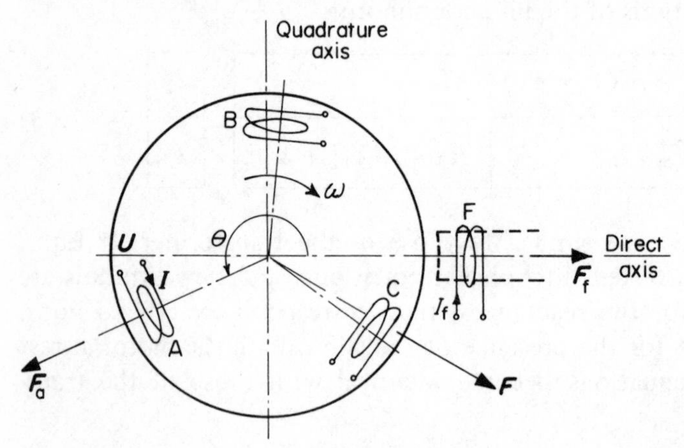

Fig. 3.5 Diagram of a synchronous generator.

Fig. 3.5, as well as Figs. 3.6 to 3.10, apply for steady operation of the machine as a generator at lagging power factor. The rotor runs at the constant synchronous speed ω_0 and it is assumed that the direction of rotation is clockwise, so that $\omega = -\omega_0$. The armature winding carries balanced polyphase currents, the phase sequence of which must be A–C–B. The clockwise rotation is chosen in order to make the angular sequence and relations in the space phasor diagram of m.m.f.s (Fig. 3.6) correspond to those in the voltage diagram (Fig. 3.7), when drawn according to the usual convention.

The m.m.f. wave due to the armature currents rotates counter clockwise relative to the armature at speed ω_0, and is therefore in a fixed position in space whatever the position of the armature. The position of the armature winding is specified by the angle θ which the axis of coil A makes with the direct axis, and in Fig. 3.5

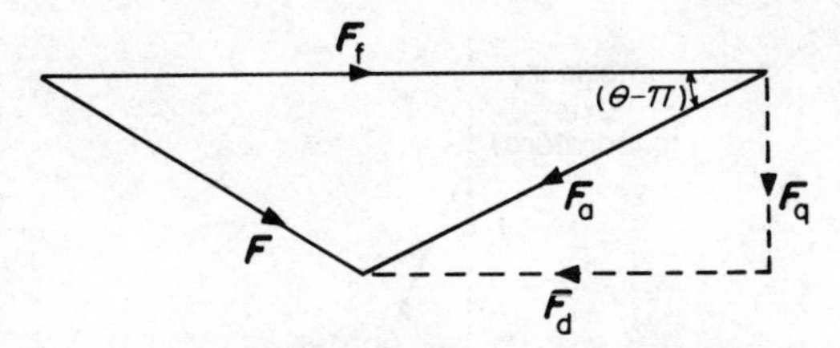

Fig. 3.6 Space phasor diagram of m.m.f.s in a synchronous generator.

the windings are shown in the position they occupy at the instant when the current in phase A is a maximum. The axis of the armature m.m.f. coincides with the axis of phase A at this instant, and hence $\boldsymbol{F}_a$, which remains fixed as the armature rotates, is in the position shown.

In the diagram, the m.m.f. phasor due to the field current $\boldsymbol{I}_f$ is along the direct axis. The resultant m.m.f. obtained by adding $\boldsymbol{F}_a$ and $\boldsymbol{F}_f$ is represented by $\boldsymbol{F}$. The space phasor diagram (Fig. 3.6) shows the relation between the three m.m.f. phasors.

The machine has a uniform air-gap, thus the resultant flux density is proportional to the m.m.f. at every point, and the flux can be considered to be the resultant of two component fluxes proportional to $\boldsymbol{F}_a$ and $\boldsymbol{F}_f$. Moreover the voltage induced in phase A is proportional to the flux, and its time phase depends on the space position of the flux, so that the voltage can be considered to be the resultant of two corresponding component voltages. Hence the space phasor diagram of m.m.fs. can be converted into the time phasor diagram of voltages given by the triangle OST in Fig. 3.7.

In Fig. 3.7, OS is the phasor $\boldsymbol{U}_0$ representing the voltage induced in phase A by the component of flux due to the field current. It is proportional to $\boldsymbol{F}_f$ and is drawn vertically because the voltage is maximum when coil A has its axis vertical. $\boldsymbol{U}_0$ is thus the open-circuit or excitation voltage, which depends only on the field current. ST represents the voltage induced by the component of flux due to the armature current $\boldsymbol{I}$, and can be considered as a reactive drop in the magnetising reactance X_{md}. It leads the current by 90° and is equal to $jX_{md}\boldsymbol{I}$. Finally, OT represents the internal voltage $\boldsymbol{U}_i$ induced by the resultant flux.

The voltage phasor diagram is completed by adding the resistance drop $R_a\boldsymbol{I}$ and the leakage reactance drop $jX_a\boldsymbol{I}$ to the

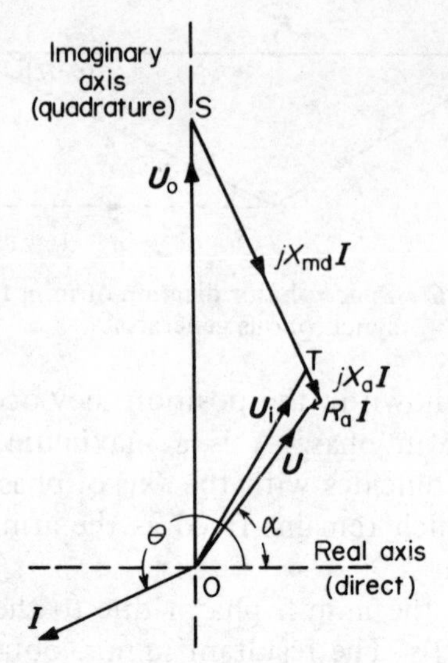

Fig. 3.7 Phasor diagram of a uniform air-gap synchronous machine.

internal voltage U_i. The resultant is equal to the terminal voltage U, which makes an angle α with the horizontal axis. Fig. 3.7 is the complete phasor diagram relating the terminal voltage U and current I.

Although the phasor diagram has been derived for the particular condition when the machine is operating as a *generator* at lagging power factor, the theory is quite general and applies for any condition of steady operation. The shape of the diagram does of course vary considerably at different loads and power factors. Alternatively the relations given by the phasor diagram may be expressed by a phasor equation or by an equivalent circuit. The equation is:

$$U = U_0 + (R_a + jX_d)I, \tag{3.4}$$

where $X_d = X_{md} + X_a$ (synchronous reactance).

The equivalent circuit of the uniform air-gap synchronous machine is shown in Fig. 3.8. It is evident that Eqn. (3.4) applies to this network. The terminal voltage (less the resistance drop) is determined as the sum of U_0, the voltage that would be induced if only I_f were present, and jX_dI the voltage that would be induced if only I were present. The result is based on the principle of superposition, which depends on the assumption of linearity.

The calculation of the currents flowing in a power system under steady conditions when the applied voltages are known is called a *load flow study.* It is possible to make such a calculation by assuming that each generator has a known excitation and including its synchronous reactance in the network. The constant field corresponds to a constant *voltage behind synchronous reactance* which is taken to be the driving voltage U_0. The value of reactance is, however, uncertain because of saturation, and moreover, it is a commoner practical condition to operate the generator with a constant terminal voltage maintained by a voltage regulator, rather than with constant excitation. Under these conditions the load flow study is purely a network problem, for which the generator voltage is assumed, and the generator characteristics do not enter into the calculations.

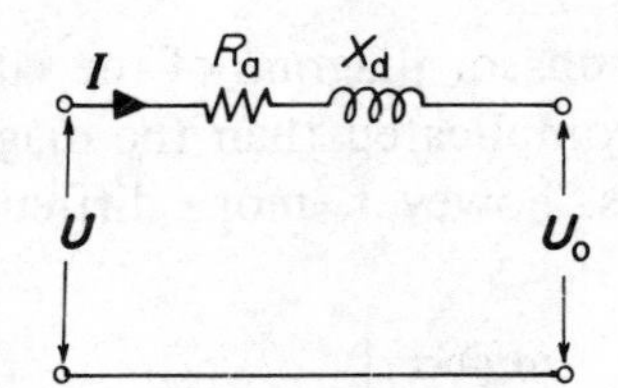

Fig. 3.8 Equivalent circuit of a uniform air-gap synchronous machine.

3.4 The salient-pole synchronous machine

For a machine with salient poles (indicated by dots in Fig. 3.5), the m.m.f. diagram in Fig. 3.6 remains unchanged. However, the flux density no longer bears the same ratio to the m.m.f. at every point, and therefore must be determined by resolving the armature m.m.f. into components $\boldsymbol{F}_d$ and $\boldsymbol{F}_q$ along the two axes. It is assumed that each component of m.m.f. produces a proportional flux along the same axis, but that the factor of proportionality is different for the two axes. The figure OSLT of Fig. 3.9 is the modified voltage phasor diagram based on this assumption.

The component phasors $\boldsymbol{F}_d$ and $\boldsymbol{F}_q$ in the space phasor diagram of m.m.f.s (Fig. 3.6) correspond to component phasors $\boldsymbol{I}_d$ and $\boldsymbol{I}_q$

in the time phasor diagram of currents (Fig. 3.9). The voltage induced by the direct-axis component of flux is represented by LT in Fig. 3.9, and is equal to $jX_{md}\boldsymbol{I}_d$, where X_{md} is the *direct-axis magnetising reactance*. Similarly the voltage induced by the quadrature-axis component of flux, represented by SL, is equal to $jX_{mq}\boldsymbol{I}_q$, where X_{mq} is the *quadrature-axis magnetizing reactance.* The open-circuit voltage $\boldsymbol{U}_0$, the internal voltage $\boldsymbol{U}_i$, the resistance and leakage reactance drops, and the terminal voltage $\boldsymbol{U}$ remain as before.

The voltage phasor equation of Fig. 3.9 is

$$\boldsymbol{U} = \boldsymbol{U}_0 + R_a\boldsymbol{I} + jX_d\boldsymbol{I}_d + jX_q\boldsymbol{I}_q, \tag{3.5}$$

where

$X_d = X_{md} + X_a$ (direct-axis synchronous reactance),

$X_q = X_{mq} + X_a$ (quadrature-axis synchronous reactance).

Thus the two-axis phasor diagram of the salient-pole machine is in itself little more complicated than the diagram of the uniform air-gap machine. It is, however, more difficult to apply, because

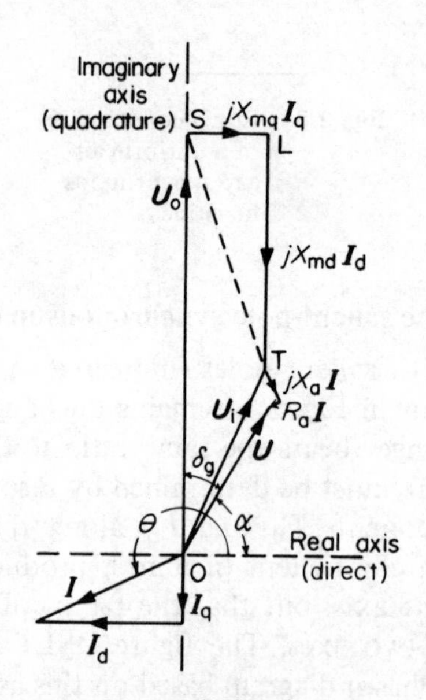

Fig. 3.9 Phasor diagram of a salient pole synchronous generator.

the resolution of the current phasor introduces the unknown angle θ. It is not possible to derive a simple single equivalent circuit for determining the current corresponding to a given supply voltage.

An approximate method for determining the angle θ and other quantities is given in the next section.

The phasor I_d is called the *direct-axis component* of the current and I_q is called the *quadrature-axis component*, because they correspond to the components of the m.m.f. along the two axes. The word 'direct' and 'quadrature' are shown on Figs. 3.7 and 3.9 in brackets, because, although they are used for the current components, they apply strictly to the space phasor diagram of m.m.f.s., and not to the time phasor diagram of currents. The resolution of the current into the axis components corresponds to the transformation from phase values to axis values in the general theory. The question is discussed further in Section 7.1 where the steady-state theory, instead of being worked out from first principles, is derived as a special case of the general theory.

3.5 Characteristic of a synchronous machine connected to an external supply

A synchronous machine normally operates in conjunction with an external supply system, and for many purposes the supply voltage may be considered as a *fixed reference*, both as regards magnitude and phase. Such a fixed supply is also called an *infinite bus*. If the machine of Fig. 3.5 is driven in such a way as to generate a fixed terminal voltage in phase A when on open circuit, the position occupied by the rotor at any instant is determined by this voltage. Conversely, the voltage of an external supply system can be associated with a uniformly rotating reference axis, whose position is that which the axis of the rotor would have if it were driven in such a way as to generate the external voltage on open circuit. The position of the rotor during any operating condition can then be measured by its displacement relative to the reference axis.

When the synchronous machine is operating steadily on load, the speed is still the constant synchronous speed ω_0, but the angular position is displaced from the reference position by a constant rotor angle δ. The angle δ is also called the *load angle*, because its value varies progressively with the load. If the rotor lags behind the reference axis, δ is positive, and the machine acts as a motor, developing motoring torque which tends to accelerate the rotor. When the rotor leads the reference axis, the machine is a generator and causes the opposite effect.

3.5.1 Power-angle relation

The curve relating the input power P and the load angle δ is called the *power-angle characteristic.* The value of δ is given by the angle between U_0 and U in the phasor diagram of Fig. 3.9, which has been drawn particularly for generator action when the value of δ is negative. Hence the angle between U and U_0 is the positive angle $\delta_g = -\delta$. Moreover the electrical input power P is negative for generator operation, and the generator output is $P_g = -P$. The relation between P_g and δ_g can be deduced from Fig. 3.9 as a simple expression if the armature resistance R_a is neglected.

Fig. 3.10 shows a simplified form of the phasor diagram of Fig. 3.9 in which $R_a I$ is now omitted, and each component of magnetizing reactance drop is combined with the corresponding component of the leakage reactance drop $jX_a I$ to give a component of synchronous reactance drop. If the symbols I, I_d and I_q

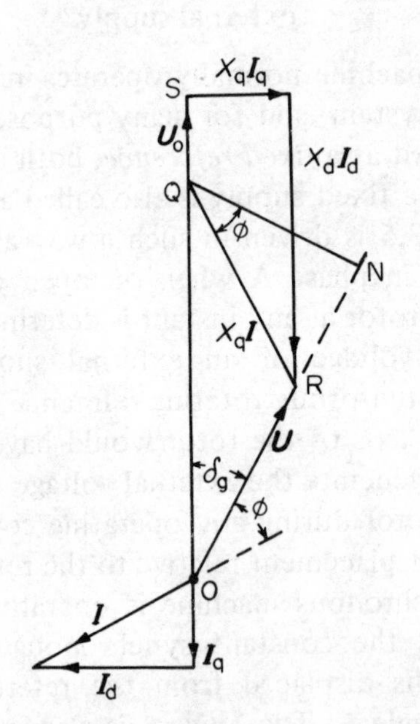

Fig. 3.10 Simplified phasor diagram of a synchronous generator.

are used for the magnitudes of the phasors representing the currents, the magnitudes of the direct and quadrature-axis components of synchronous reactance drop are $X_d I_d$ and $X_q I_q$. Thus Fig. 3.10 agrees with Eqn. (3.5) if the resistance drop is omitted.

In order to determine δ_g, a line RQ is drawn perpendicular to the current phasor, meeting OS at Q, and a perpendicular QN is dropped on OR produced. The angle RQN = ϕ, and

$$\mathrm{RQ} = \frac{X_q I_q}{\sin \mathrm{OQR}} = X_q I.$$

Hence

$$\tan \delta_g = \frac{\mathrm{NQ}}{\mathrm{ON}} = \frac{X_q I \cos \phi}{U + X_q I \sin \phi}. \tag{3.6}$$

Also

$$\left.\begin{aligned} I \cos \phi &= I_d \sin \delta_g + I_q \cos \delta_g \\ I \sin \phi &= I_d \cos \delta_g - I_q \sin \delta_g \\ U_0 - U \cos \delta_g &= I_d X_d \\ U \sin \delta_g &= I_q X_q \end{aligned}\right\} \tag{3.7}$$

The generator output power is given by

$$P_g = UI \cos \phi,$$

which can be written, using Eqns. (3.7),

$$P_g = \frac{UU_0}{X_d} \sin \delta_g + \frac{U^2}{2}\left(\frac{1}{X_q} - \frac{1}{X_d}\right) \sin 2\delta_g \tag{3.8}$$

The full-line curve in Fig. 3.11 shows the power-angle characteristic obtained by plotting P_g against δ_g for a salient-pole machine. For a uniform air-gap machine, with $X_d = X_q$, the second term in Eqn. (3.8) is zero, and the curve becomes simply the sine curve shown by the dotted line.

3.5.2 Synchronizing torque coefficient

Since the torque of a synchronous machine, with fixed terminal voltage and field current, increases with the load angle, the machine is equivalent to a torsional spring connected between the

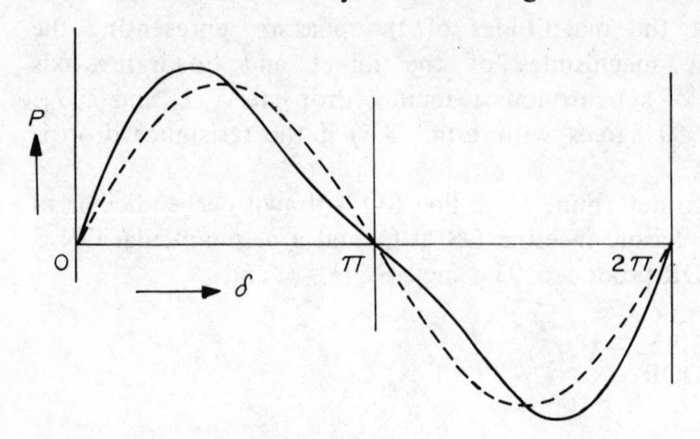

Fig. 3.11 Power-angle characteristic of a synchronous machine.

output shaft and the rotating reference axis which depends on the supply voltage, as explained on p. 53.

At small load angles, whether positive or negative, the torque is approximately proportional to the displacement, but as the angle increases, the torque increases less rapidly and eventually reaches a maximum. Thus the spring, to which the machine is equivalent, has a non-linear characteristic. The effective spring constant for small changes relative to any steady condition is given by the slope of the curve. It is called the *synchronizing torque coefficient*, and is denoted by P_0. Its value is:

$$P_0 = \frac{dP_g}{d\delta_g} = \frac{UU_0}{X_d}\cos\delta_g + U^2\left(\frac{1}{X_q} - \frac{1}{X_d}\right)\cos 2\delta_g. \tag{3.9}$$

The effective spring constant, deduced from the steady-state characteristic, is only correct for slow changes. When rapid changes occur the conditions may be greatly modified by the action of the field and damper windings, as shown in Section 7.5.

3.5.3 Operating chart of a synchronous generator [9]

Any condition of steady operation of a uniform air-gap synchronous machine in parallel with a fixed supply voltage U (or infinite bus) is defined by any two of the three quantities, active power P, reactive power Q and load angle δ (dropping the suffix g). A typical steady operating point is indicated by Z in the diagram of Fig. 3.12. The diagram is called an *operating chart* and

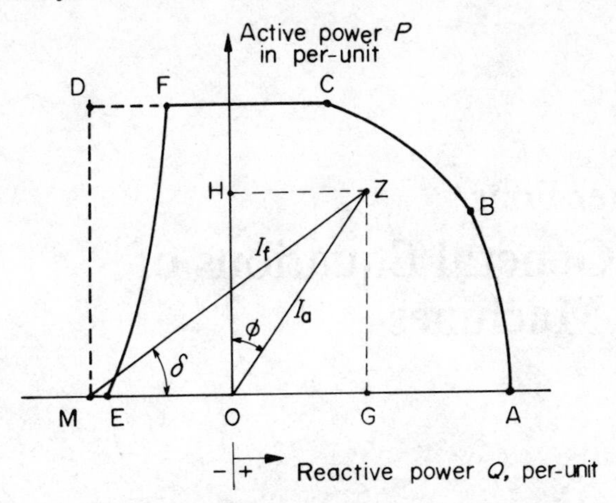

Fig. 3.12 Operating chart for a synchronous generator showing the safe operating region.

its derivation is based on the phasor diagram of Fig. 3.7 with armature resistance neglected. The sign convention used is that, at the terminals of an over-excited synchronous generator, reactive power is defined as positive.

Fig. 3.12, which applies to a uniform air-gap machine, shows a triangular current phasor diagram OZM. If saturation is neglected, OZ is proportional to I_a, MZ to I_f, and OM to U^2/X_d, which is a measure of the short-circuit ratio.

The safe operating region on the P–Q chart is limited by the following consideration: curve AB is part of a circle with centre M and is determined by the maximum safe field current; curve BC is part of a circle around O and is determined by the maximum safe armature current; the horizontal line CD is determined by the maximum safe prime mover output while the vertical line MD is the theoretical steady-state stability boundary occurring, in a uniform air-gap machine, when $\delta = 90$ degrees. In practice, Z is not permitted to move as far left as line MD, because the latter represents a condition of instability where the smallest increment in load causes the machine to fall out of step. Hence a safety margin between MD and curve EF is allowed. The area enclosed by points ABCFE is called the *safe operating area.*

A similar chart based on Fig. 3.9 can be drawn for the salient pole machine [18].

Chapter Four

The General Equations of A.C. Machines

4.1 Equations in terms of phase variables

4.1.1 Three-phase machine

Any machine, other than those with commutators, consists of a set of coils for which voltage equations can be written in terms of self and mutual inductances and resistances. The machine however differs from a static circuit in that the coils are in relative motion and consequently that some of the inductances are functions of the relative position. The idealized synchronous machine of Fig. 1.3 has six coils and there are six voltage equations, in which many of the coefficients are periodic functions of the angle θ.

The self-inductances L_{ff}, L_{kkd}, L_{kkq} of the field and damper coils and the mutual inductance L_{fkd}, between F and KD, are all constants. There is no mutual inductance between KQ and F or KD. The armature self-inductances L_{aa}, L_{bb}, L_{cc} of a machine with salient poles vary periodically with period π between a maximum when the coil is on a pole axis and a minimum when it is on an interpolar axis. Thus

$$L_{aa} = A_0 + A_2 \cos 2\theta + A_4 \cos 4\theta + \ldots$$

L_{bb} and L_{cc} are given by the same expression with θ replaced by $(\theta - 2\pi/3)$ and $(\theta - 4\pi/3)$ respectively.

The armature mutual inductances L_{bc}, L_{ca} and L_{ab} vary periodically with period π. L_{bc} is a maximum when A is on the quadrature axis, hence

$$L_{bc} = -B_0 + B_2 \cos 2\theta + B_4 \cos 4\theta + \ldots$$

L_{ca} and L_{ab} are given by the same expression with θ replaced by $(\theta - 2\pi/3)$ and $(\theta - 4\pi/3)$ respectively.

The mutual inductances between an armature coil and a field or damper coil vary periodically with period 2π. L_{af} is a maximum when A is on the direct axis and zero when A is on the quadrature axis. Thus

$$L_{af} = C_1 \cos \theta + C_3 \cos 3\theta + \ldots$$

$$L_{akd} = D_1 \cos \theta + D_3 \cos 3\theta + \ldots$$

$$L_{akq} = D_1 \sin \theta - D_3 \sin 3\theta + \ldots$$

The remaining inductances involving B and C are obtained by replacing θ by $(\theta - 2\pi/3)$ and $(\theta - 4\pi/3)$ in the appropriate expressions.

The equations of the machine can be written down for any possible arrangement of the coils and the magnetic circuit. The only assumptions made are that there is no saturation and that there is no distributed conducting material in which eddy currents can flow. No assumption is made about the shape of the air-gap m.m.f. wave set up by any winding. The coefficients in the Fourier series can be calculated from the design or determined by test. It is evident however that the determination of the parameters and the solution of equations in this form is a complicated process, but is quite possible with a digital computer.

For many purposes the equations can be simplified by neglecting the Fourier terms of order 3 and higher, thus assuming that all of the inductances vary sinusoidally, with an additional constant term in some cases. The equations relating the terminal voltages to the currents in the six circuits of Fig. 1.3 are then expressed by the matrix equation

$$\boldsymbol{u} = \boldsymbol{Zi} \tag{4.1}$$

where $\boldsymbol{Z}$ is given by Eqn. (4.2).

4.1.2 Two-phase machine

For a two-phase machine having two armature phases α and β located at 90 degrees, as shown in Fig. 4.1, the self-inductances of

$\mathbf{Z} =$	a	b	c	f	kd	kq
a	$R_a + p(A_0 + A_2 \cos 2\theta)$	$p\left[-B_0 + B_2 \cos\left(2\theta - \frac{2\pi}{3}\right)\right]$	$p\left[-B_0 + B \cos\left(2\theta - \frac{4\pi}{3}\right)\right]$	$pC_1 \cos\theta$	$pD_1 \cos\theta$	$pD_1 \sin\theta$
b	$p\left[-B_0 + B_2 \cos\left(2\theta - \frac{2\pi}{3}\right)\right]$	$R_a + p\left[A_0 + A_2 \cos\left(2\theta - \frac{4\pi}{3}\right)\right]$	$p(-B_0 + B_2 \cos 2\theta)$	$pC_1 \cos\left(\theta - \frac{2\pi}{3}\right)$	$pD_1 \cos\left(\theta - \frac{2\pi}{3}\right)$	$pD_1 \sin\left(\theta - \frac{2\pi}{3}\right)$
c	$p\left[-B_0 + B_2 \cos\left(2\theta - \frac{4\pi}{3}\right)\right]$	$p(-B_0 + B_2 \cos 2\theta)$	$R_a + p\left[A_0 + A_2 \cos\left(2\theta - \frac{2\pi}{3}\right)\right]$	$pC_1 \cos\left(\theta - \frac{4\pi}{3}\right)$	$pD_1 \cos\left(\theta - \frac{4\pi}{3}\right)$	$pD_1 \sin\left(\theta - \frac{4\pi}{3}\right)$
f	$pC_1 \cos\theta$	$pC_1 \cos\left(\theta - \frac{2\pi}{3}\right)$	$pC_1 \cos\left(\theta - \frac{4\pi}{3}\right)$	$R_f + L_{ff}p$	$L_{fkd}p$	
kd	$pD_1 \cos\theta$	$pD_1 \cos\left(\theta - \frac{2\pi}{3}\right)$	$pD_1 \cos\left(\theta - \frac{4\pi}{3}\right)$	$L_{fkd}p$	$R_{kd} + L_{kkd}p$	
kq	$pD_1 \sin\theta$	$pD_1 \sin\left(\theta - \frac{2\pi}{3}\right)$	$pD_1 \sin\left(\theta - \frac{4\pi}{3}\right)$			$R_{kq} + L_{kkq}p$

(4.2)

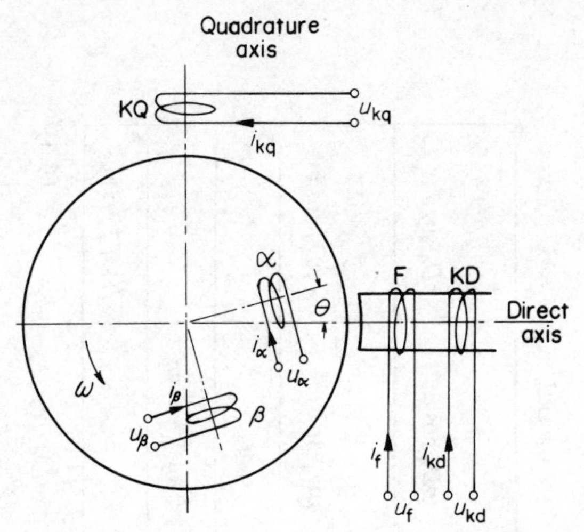

Fig. 4.1 Diagram of an idealized two-phase machine.

the field and damper coils and the mutual inductance between F and KD are constant as before [26].

The inductances of a salient-pole machine vary periodically as follows when the Fourier terms of order 3 and higher are neglected:–

$$L_{\alpha\alpha} = A_0 + A_2 \cos 2\theta$$
$$L_{\alpha\beta} = B_2 \sin 2\theta$$
$$L_{\alpha f} = C_1 \cos \theta$$
$$L_{\alpha kd} = D_1 \cos \theta$$
$$L_{\alpha kq} = D_1 \sin \theta$$

The remaining inductances involving coil β are obtained by replacing θ by $(\theta - \pi/2)$ in the appropriate expressions.

The equations for the terminal voltage of the five coils in Fig. 4.1 are expressed by Eqn. (4.1), where $\boldsymbol{Z}$ is given by Eqn. (4.3).

4.2 Transformation between various reference frames

4.2.1 The three-phase to two-axis transformation due to Park [4]

It has been noted that the Eqns. (4.1) relating the three-phase voltages and currents, where many of the coefficients are functions of θ, are difficult to handle directly. Subject to some

$$
\mathbf{Z} = \begin{array}{c|c|c|c|c|c|}
 & \alpha & \beta & f & kd & kq \\
\hline
\alpha & R_a + p(A_0 + A_2 \cos 2\theta) & pB_2 \sin 2\theta & pC_1 \cos \theta & pD_1 \cos \theta & pD_1 \sin \theta \\
\hline
\beta & pB_2 \sin 2\theta & R_\alpha + p(A_0 - A_2 \cos 2\theta) & pC_1 \sin \theta & pD_1 \sin \theta & -pD_1 \cos \theta \\
\hline
f & pC_1 \cos \theta & pC_1 \sin \theta & R_f + L_{ff}p & L_{fkd}p & \\
\hline
kd & pD_1 \cos \theta & pD_1 \sin \theta & L_{fkd}p & R_{kd} + L_{kkd}p & \\
\hline
kq & pD_1 \sin \theta & -pD_1 \cos \theta & & & R_{kq} + L_{kkq}p \\
\hline
\end{array}
\tag{4.3}
$$

assumptions, which are discussed in the following pages, the solution becomes much easier if the armature variables are transformed to new variables related to a reference frame fixed to the field system.

The direct and quadrature axis currents i_d, i_q are defined as the currents in fictitious coils, located on the axes and each having the same number of turns as a phase coil, which would set up the same m.m.f. wave as the actual currents i_a, i_b, i_c. Because the three actual coils are replaced by a system of two axis coils, it is advantageous to change the base value of current in the axis coils to a value one and a half times the base current in the phase coils. It is shown later that as a result of this choice of base values, and of the transformation Eqns. (4.4) and (4.6), which define the new variables, the circuit Eqns. (4.15) relating axis quantities are obtained in a form which does not introduce a numerical factor.

Assuming that the m.m.f. wave due to the current i_a in the phase coil A of Fig. 1.3 is sinusoidal, the maximum m.m.f. is proportional to i_a and occurs at the axis of the coil; that is, at the angular position θ. The m.m.f. wave due to i_a may be resolved into two components, one along each of the direct and quadrature axes. The amplitude of the direct-axis component is:

$$k_m i_a \cos\theta$$

where k_m is a constant.

The direct-axis component of the resultant m.m.f. wave, due to the combined action of the three-phase currents, is therefore of amplitude

$$k_m \left\{ i_a \cos\theta + i_b \cos\left(\theta - \frac{2\pi}{3}\right) + i_c \cos\left(\theta - \frac{4\pi}{3}\right)\right\}$$

The amplitude of the m.m.f. wave due to the current i_d in the direct-axis coil D of Fig. 4.2, taking into account the change of unit explained above, is

$$\frac{3}{2} k_m i_d .$$

Hence i_d is given by

$$i_d = \frac{2}{3} \left\{ i_a \cos\theta + i_b \cos\left(\theta - \frac{2\pi}{3}\right) + i_c \cos\left(\theta - \frac{4\pi}{3}\right)\right\}$$

similarly

$$i_q = \frac{2}{3}\left\{i_a \sin\theta + i_b \sin\left(\theta - \frac{2\pi}{3}\right) + i_c \sin\left(\theta - \frac{4\pi}{3}\right)\right\}$$

The zero-sequence current

The current i_z is defined by: $i_z = \frac{1}{3}(i_a + i_b + i_c)$.

The term 'zero-sequence current' is adopted from the analogy with the 'zero-sequence component' used in symmetrical component theory, but i_z is an instantaneous value of current, which may vary with time in any manner. It may be visualized physically as the magnitude of each of a set of equal currents, flowing in all three phases and therefore producing no resultant air-gap m.m.f.

The current transformations

The transformation equations giving the new currents i_d, i_q, i_z, in terms of the actual currents i_a, i_b, i_c, are therefore expressed by the following matrix equation:

$$\begin{bmatrix} i_d \\ i_q \\ i_z \end{bmatrix} = \frac{2}{3}\begin{bmatrix} \cos\theta & \cos\left(\theta - \frac{2\pi}{3}\right) & \cos\left(\theta - \frac{4\pi}{3}\right) \\ \sin\theta & \sin\left(\theta - \frac{2\pi}{3}\right) & \sin\left(\theta - \frac{4\pi}{3}\right) \\ \frac{1}{2} & \frac{1}{2} & \frac{1}{2} \end{bmatrix}\begin{bmatrix} i_a \\ i_b \\ i_c \end{bmatrix} \quad (4.4)$$

The equations of the *inverse transformation*, giving the actual currents in terms of the new currents, are obtained by solving the above equations. Hence

$$\begin{bmatrix} i_a \\ i_b \\ i_c \end{bmatrix} = \begin{bmatrix} \cos\theta & \sin\theta & 1 \\ \cos\left(\theta - \frac{2\pi}{3}\right) & \sin\left(\theta - \frac{2\pi}{3}\right) & 1 \\ \cos\left(\theta - \frac{4\pi}{3}\right) & \sin\left(\theta - \frac{4\pi}{3}\right) & 1 \end{bmatrix}\begin{bmatrix} i_d \\ i_q \\ i_z \end{bmatrix} \quad (4.5)$$

The voltage transformations

The new voltages are defined by a set of equations exactly similar to those for the currents:

$$\begin{bmatrix} u_d \\ u_q \\ u_z \end{bmatrix} = \frac{2}{3} \begin{bmatrix} \cos\theta & \cos\left(\theta - \frac{2\pi}{3}\right) & \cos\left(\theta - \frac{4\pi}{3}\right) \\ \sin\theta & \sin\left(\theta - \frac{2\pi}{3}\right) & \sin\left(\theta - \frac{4\pi}{3}\right) \\ \frac{1}{2} & \frac{1}{2} & \frac{1}{2} \end{bmatrix} \begin{bmatrix} u_a \\ u_b \\ u_c \end{bmatrix} \quad (4.6)$$

The equations of the inverse transformation are:

$$\begin{bmatrix} u_a \\ u_b \\ u_c \end{bmatrix} = \begin{bmatrix} \cos\theta & \sin\theta & 1 \\ \cos\left(\theta - \frac{2\pi}{3}\right) & \sin\left(\theta - \frac{2\pi}{3}\right) & 1 \\ \cos\left(\theta - \frac{4\pi}{3}\right) & \sin\left(\theta - \frac{4\pi}{3}\right) & 1 \end{bmatrix} \cdot \begin{bmatrix} u_d \\ u_q \\ u_z \end{bmatrix} \quad (4.7)$$

Power input

Using the definition of base power given on p. 20, the power input to a three-phase armature winding is given by:

$$P = \tfrac{1}{3}(u_a i_a + u_b i_b + u_c i_c). \quad (4.8)$$

By substituting the expressions for the currents and voltages given by the transformation Eqns. (4.5) and (4.7) it can readily be shown that:

$$P = \tfrac{1}{2}(u_d i_d + u_q i_q) + u_z i_z \quad (4.9)$$

For normal steady operation with balanced polyphase voltage and currents, the quantities u_z and i_z are zero. Hence Eqn. (4.9) accords with the definition of unit power for a machine with two main circuits (see p. 20 and Section 13.2). Comparison of

Eqns. (4.8) and (4.9) shows that the power is non-invariant since the factors 1/3 and 1/2 are unequal.

The transformation can, if desired, be regarded as a purely mathematical process which transforms the description of the voltages and currents from one reference frame to another. However, the physical interpretation in terms of the field axes is helpful in relating the a.c. machine to the primitive machine discussed in Chapter 2. The choice of the voltage transformation Eqn. (4.6) is made arbitrarily, but this is later given a physical meaning.

4.2.2 Other transformations

Three reference frames for the armature coils have been considered so far, viz. three phases (a,b,c), two phases (α,β) and two axes (d,q). It is possible to transform a three-phase winding to an equivalent two-phase winding but it is necessary to include a zero sequence component in addition. If coils α and A are in the same position, the transformation matrix from frame (a,b,c) to (α,β,z) is obtained by putting $\theta = 0$ in Eqn. (4.4) to yield

$$\begin{bmatrix} i_\alpha \\ i_\beta \\ i_z \end{bmatrix} = \frac{2}{3} \begin{bmatrix} 1 & -\frac{1}{2} & -\frac{1}{2} \\ 0 & \sqrt{\frac{3}{2}} & -\sqrt{\frac{3}{2}} \\ \frac{1}{2} & \frac{1}{2} & \frac{1}{2} \end{bmatrix} \begin{bmatrix} i_a \\ i_b \\ i_c \end{bmatrix} \tag{4.10}$$

The three-phase Eqns. (4.2) can be transformed to the two-phase Eqns. (4.3) by using the above transformation, together with the corresponding voltage transformation, and ignoring the zero sequence component.

The transformation from the (α,β) reference frame to the (d,q) frame is much simpler than that for the three-phase machine.

$$\begin{bmatrix} i_d \\ i_q \end{bmatrix} = \begin{bmatrix} \cos\theta & \sin\theta \\ \sin\theta & -\cos\theta \end{bmatrix} \begin{bmatrix} i_\alpha \\ i_\beta \end{bmatrix} \tag{4.11}$$

where θ is the angle between the axes of the α and D coils. The transformation equation (4.4) can be obtained by combining Eqns. (4.10) and (4.11) by matrix multiplication. For this purpose an additional row and column for i_z must be added to Eqn. (4.11).

A further reference frame, designated the (D,Q) frame, which rotates at synchronous speed relative to the armature winding, is defined and explained in Section 9.5.

4.2.3 The nature of the assumptions in the two-axis equations

By using the transformations given above, the Eqns. (4.1) can be converted into new equations in terms of the axis variables. The operational impedances Z in Eqns. (4.2) and (4.3) are transformed by calculating the matrix product $Z' = C^T ZC$ with a factor $\frac{3}{2}$ in the three-phase case. In many books on the subject, and indeed in Park's original paper, the two-axis equations are derived by first setting down the phase equations and then carrying out the transformation. However, since the derivative operator p must act on terms like $A_2 \cos 2\theta \cdot i_a$, where $i_a = i_d \cos\theta + i_q \sin\theta + i_z$, the working out of the transformation is a formidable piece of algebraic manipulation and many writers have preferred to limit their treatment to a machine with no damper winding and often also to the two-phase case.

In the present treatment, the direct derivation of the two-axis equations explained in Section 4.3 is preferred, because it shows more clearly the physical meaning of the equations and the assumptions made. Moreover the derivation is simpler and is readily applied to the complete three-phase machine with damper windings.

The simple form of the axis equations obtained by transforming the phase equations is a result of the following assumptions.

1. That the terms of order 3 and higher in the Fourier expressions for the inductances can be neglected.
2. $A_2 = B_2$ in Eqns. (4.2) and (4.3).

Alternatively the assumption underlying Park's equations, as discussed on p. 23, is that the armature winding is sinusoidally distributed, so that a current in the winding does not produce any harmonic m.m.f.s and a harmonic flux does not induce any voltage in the winding. It is shown below that the two assumptions are equivalent.

In order to explain the nature of the assumptions in a simple manner a simplified machine with a two-phase armature winding and no damper winding is used as an example. The impedance matrix is,

$$\begin{array}{c|c|c|c|}
 & \alpha & \beta & f \\
\hline
\alpha & R_a + p(A_0 + A_2 \cos 2\theta) & pB_2 \sin 2\theta & pC_1 \cos \theta \\
\hline
\beta & pB_2 \sin 2\theta & R_\alpha + p(A_0 - A_2 \cos 2\theta) & pC_1 \sin \theta \\
\hline
f & pC_1 \cos \theta & pC_1 \sin \theta & R_f + L_{ff}p \\
\hline
\end{array} \quad (4.12)$$

The transformation is

$$\begin{bmatrix} i_d \\ i_q \\ i_f \end{bmatrix} = \begin{bmatrix} \cos\theta & \sin\theta & \\ \sin\theta & -\cos\theta & \\ & & 1 \end{bmatrix} \begin{bmatrix} i_\alpha \\ i_\beta \\ i_f \end{bmatrix} \quad (4.13)$$

The assumption that $A_2 = B_2$ is important. It has been shown in [26] that, if A_2 and B_2 differ by an amount L_c, the transformed equations contain additional terms in $L_c \cos 4\theta$ and $L_c \sin 4\theta$ and would be no easier to handle than the original equations.

The assumption that $A_2 = B_2$ has been justified by basing it on the more fundamental assumption that the permeance wave in the salient-pole machine is sinusoidal [23]. The permeance at a point in the air-gap is defined as the ratio of flux density to m.m.f. at the point. While the assumption is reasonable at points under the pole face, it is doubtful at points in the interpolar space of a salient-pole machine. Indeed the concept of a permeance is itself imprecise, since, as shown in [5], a different permeance curve is obtained, if the m.m.f. is sinusoidally distributed, from that obtained on the usual assumption that the magnetic potential at each surface is constant.

A better basis for the assumption that $A_2 = B_2$ is to start with the essential assumption stated above, namely that the armature winding is sinusoidally distributed. A further assumption is that the machine construction is symmetrical with respect to both direct and quadrature axes. In the two-phase machine a current i_α in phase α at angle θ produces m.m.f. components $k_m i_\alpha \cos\theta$ and $k_m i_\alpha \sin\theta$ on the axes. The m.m.f. components produce axis components of air-gap flux, which depend on the magnetising inductances L_{md} and L_{mq} (defined on p. 80), so that the resulting flux linkage with winding α is

$$(L_{md}\cos^2\theta + L_{mq}\sin^2\theta)i_\alpha = \left[\frac{L_{md}+L_{mq}}{2} + \frac{L_{md}-L_{mq}}{2}\cos 2\theta\right] i_d \tag{4.14}$$

The resultant flux linkage with winding β at angle $(\theta - \pi/2)$ is

$$i_\alpha(L_{md} - L_{mq})\cos\theta\,.\,\sin\theta = \left[\frac{L_{md}-L_{mq}}{2}\,.\,\sin 2\theta\right] i_\alpha$$

Hence the two coefficients A_2 and B_2 are both equal to

$$\left[\frac{L_{md}-L_{mq}}{2}\right]$$

and there are no higher terms in the Fourier series.

Application of the two-axis equations to the analysis of normal synchronous and induction machines, which constitute the great majority of electrical machines in use, gives very good results. They form the basis of the established methods of power system analysis, where it is affected by the rotating machines. That this is so, even for salient pole machines, depends on the fact that the design procedure, for example, the use of short-pitch windings, is such that the winding harmonics are made as small as possible. It is not that the flux wave is necessarily sinusoidal, as it obviously is not when a sinusoidal m.m.f. wave is applied to the quadrature axis of a salient-pole machine. The assumption of a sinusoidal permeance wave is unrealistic for such a condition. The reason why the two-axis theory gives good results is that the winding only responds to the fundamental flux.

When the saliency is extreme, as in the reluctance motor, or when the harmonics are large, as in some types of change-pole motor, further consideration may be necessary. Some examples of conditions for which the two-axis assumptions do not apply, are discussed in Chapter 12.

4.3 Direct derivation of two-axis equations

4.3.1 Flux linkage

The base value of flux in the per-unit system is defined on p. 12 for a transformer to be such that unit rate of change of flux induces base voltage in any coil. For a machine the definition must be amplified by the stipulation that the axis of the flux wave is coincident with the axis of the coil in which the voltage is induced. For this condition the flux becomes identical with the 'flux linkage' with the coil. In the theory that follows the flux linkage is denoted by the symbol ψ and is used as a measure of the flux.

At any instant the sinusoidal flux wave existing in the air-gap requires a magnitude and an angle to define it completely. It may therefore be expressed by the flux linkage ψ_m with a coil having the same axis, together with an angle giving the angular position of the axis. Alternatively the flux wave may be resolved into two components on the direct and quadrature axes, and expressed by the corresponding direct and quadrature axis flux linkages ψ_{md} and ψ_{mq}. In practice the resolution into axis components proves to be the best method of expressing the instantaneous value of the flux.

Following Park's terminology, each of the flux linkages ψ_{md} and ψ_{mq} used to express the air-gap flux is understood to be the flux linkage with a coil on the corresponding axis. Subject to this condition, they represent the flux linkage with any of the various coils whatever the actual number of turns, because of the way in which the base value of voltage in any secondary winding is defined, as explained on p. 12.

4.3.2 The induced voltage in the armature

Assuming that the component flux linkages ψ_{md} and ψ_{mq} are known, the induced voltage in an armature coil can be determined and a voltage equation for the circuit can be derived. The voltage induced in coil A depends on its angular position θ as well as on the flux linkage components, since the voltage induced in a coil is reduced when the axis of the coil is displaced from the axis of the flux. The voltage induced by ψ_{md} is $-p(\psi_{md} \cos \theta)$ and that due to ψ_{mq} is $-p(\psi_{mq} \sin \theta)$. Hence the total internal voltage, opposing the voltage induced by the main air-gap flux, is

$$p(\psi_{md} \cos \theta + \psi_{mq} \sin \theta)$$

Now the air-gap flux is produced by the combined action of the currents in all the windings on the stator and rotor. The armature currents, however, also produce local fluxes which do not cross the air-gap but which nevertheless link phase A. The current i_a produces a leakage reactance drop $L_1 p i_a$, where L_1 is the leakage inductance of coil A. Currents i_b and i_c produce drops $-L_m p i_b$ and $-L_m p i_c$ in coil A, where L_m is the part of the mutual inductance between two armature coils due to flux which does not cross the air-gap. The terms have negative signs because the coils are displaced from each other by 120°. To a good approximation the inductances L_1 and L_m are independent of rotor position.

The impressed voltage u_a is equal to the sum of the internal voltage due to the main air-gap flux, the drops due to the local armature fluxes, and the resistance drop $R_a i_a$. Hence the equation is:

$$u_a = p(\psi_{md} \cos\theta + \psi_{mq} \sin\theta) + (R_a + L_1 p) i_a - L_m p i_b - L_m p i_c.$$

Using the relation $i_a + i_b + i_c = 3i_z$, and the value of i_a given by Eqns. (4.5), the equation becomes:

$$\begin{aligned} u_a &= p[\psi_{md} \cos\theta + \psi_{mq} \sin\theta] \\ &\quad + (L_1 + L_m) p(i_d \cos\theta + i_q \sin\theta + i_z) - 3L_m p i_z + R_a i_a \\ &= p[(\psi_{md} + L_a i_d) \cos\theta + (\psi_{mq} + L_a i_q) \sin\theta] + L_z p i_z + R_a i_a \\ &= p[\psi_d \cos\theta + \psi_q \sin\theta] + L_z p i_z + R_a i_a, \end{aligned}$$

where the following new quantities are introduced:

$$\begin{aligned} L_a &= L_1 + L_m, \\ L_z &= L_a - 3L_m, \\ \psi_d &= \psi_{md} + L_a i_d, \\ \psi_q &= \psi_{mq} + L_a i_q. \end{aligned}$$

The quantities ψ_d and ψ_q are the total flux linkages with an armature coil located on the appropriate axis, due to both the main air-gap flux and the armature leakage flux; L_a is the 'effective' leakage inductance of either of the axis coils. The inductance L_z associated with the zero-sequence current is the *zero-sequence inductance*, which is a well-known quantity in symmetrical component theory.

Equations for the voltage u_d, u_q, u_z may now be found by expanding and rearranging the above equation, using the values of

i_a and u_a from the transformation equations (4.5) and (4.7), differentiating the product terms in the square bracket and putting $p\theta = \omega$. Thus

$$(u_d - R_a i_d)\cos\theta + (u_q - R_a i_q)\sin\theta + (u_z - R_a i_z - L_z p i_z)$$
$$= \cos\theta\,(p\psi_d + \omega\psi_q) + \sin\theta\,(p\psi_q - \omega\psi_d).$$

This equation must hold for all values of θ and it therefore follows by equating coefficients that:

$$\left.\begin{aligned} u_d &= p\psi_d + \omega\psi_q + R_a i_d \\ u_q &= -\omega\psi_d + p\psi_q + R_a i_q \\ u_z &= (R_a + L_z p) i_z \end{aligned}\right\} \tag{4.15}$$

The method of equating coefficients, which is used frequently throughout the book, provides an elegant method of obtaining the above result. The method is a rigorous one, but any reader who is not satisfied about its validity should work through the rather lengthy process of obtaining expressions for u_a, u_b and u_c, and substituting in the transformation equations to obtain u_d, u_q and u_z, as explained on p. 67. It would lead to the same result.

Eqns. (4.15) form the basis of the two-axis theory. The quantities u_z and i_z are related to each other independently of the others, and for many problems only the axis voltages and currents need to be considered. Thus the transformation brings about a great simplification.

4.3.3 The fictitious axis coils in the primitive machine

The two equations for the axis quantities in Eqn. (4.15) are identical with Eqns. (2.5) for the d.c. machine. It follows that, for the a.c. machine, the voltages u_d and u_q, which were defined arbitrarily by Eqns. (4.6), can be interpreted as the impressed voltages on the direct and quadrature axis coils, provided that these coils possess the pseudo-stationary property defined on p. 9. The synchronous machine of Fig. 1.3 can then be replaced by the primitive machine of Fig. 4.2, in which the coils D and Q are pseudo-stationary coils located on the axes. Thus the same primitive machine diagram applies to both the d.c. machine and the a.c. machine and the primitive equations can be regarded as generalized equations applicable to all the principal machine types.

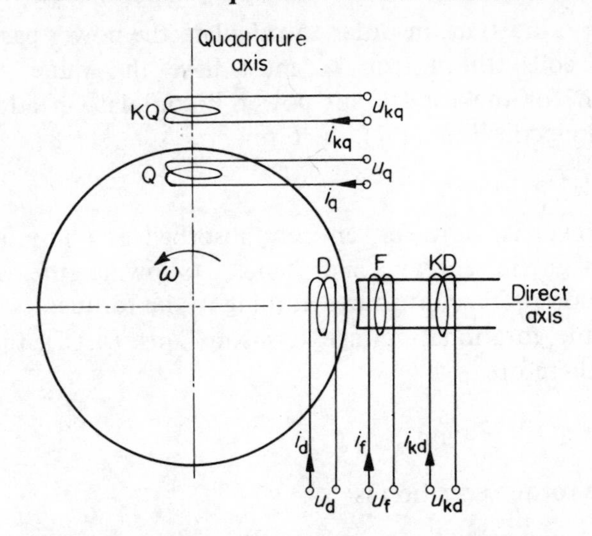

Fig. 4.2 Diagram of a synchronous machine with two damper coils.

As with the d.c. machine the terms in Eqns. (4.15) containing p are called transformer voltages and those containing ω are called rotational voltages.

The form of the armature voltage equations leads to a simple method of writing down the complete equations of a synchronous machine. If the coils D and Q of Fig. 4.2 were ordinary stationary coils the voltage equations would be:

$$u_d = p\psi_d + R_a i_d$$
$$u_q = p\psi_q + R_a i_q.$$

Eqns. (4.15) differ from these equations only by the presence of the rotational voltage terms. The complete equations for the machine of Fig. 4.2 can therefore be derived by first writing down the five equations for a set of five stationary coils, and then adding to the two armature equations additional terms for the rotational voltages. The method is used in the following two sections.

4.3.4 The torque equation

It is shown on p. 21 that the electrical torque developed by a machine, which has two armature coils D and Q on the direct and quadrature axis as in Fig. 1.5, may be determined from the rotational voltage terms in the armature equations. It has also been shown on p. 65 that, in order to calculate the power passing into the axis coils, the factor k_p must have the value 1/2. The expression for the total input power P contains, in addition to those included in Eqn. (1.7), the term

$$u_z i_z = R_a i_z^2 + L_z i_z p i_z$$

This power is, however, entirely absorbed as ohmic loss or as stored magnetic energy, and hence, following the argument explained on p. 21, contributes nothing to the torque.

From the rotational voltage terms in Eqns. (4.15), the output power is therefore:

$$P_e = \frac{\omega}{2} (\psi_q i_d - \psi_d i_q).$$

Hence the torque equation is:

$$M_e = \frac{\omega_0}{2} (\psi_d i_q - \psi_q i_d). \tag{4.16}$$

The expression for the torque shows the two components of torque discussed on p. 18. The first component is due to the interaction between the direct-axis flux and the quadrature axis current and has a positive sign. The second component is due to the interaction between the quadrature-axis flux and the direct-axis current and has a negative sign. No torque is produced by interaction between flux and current on the same axis.

4.3.5 Field and damper windings

The field winding of a synchronous machine consists of coils round the poles and forms a single circuit. In the diagram of the idealized two-pole machine it is represented by a single coil F on the direct axis, as in Fig. 4.2 or 4.3. Under normal conditions the impressed voltage u_f is supplied from a d.c. source.

The damper winding accounts for all the other closed circuits on the field system. The commonest form of damper winding consists of squirrel-cage bars in the pole face connected together at the ends by rings or segments. Sometimes a field collar, provided as a support for the field winding, forms a closed circuit round the pole. There are also eddy current paths in the iron, whether solid or laminated, of the magnet system. All these circuits should be allowed for in a complete representation. Normally no external voltages are impressed on the damper circuits.

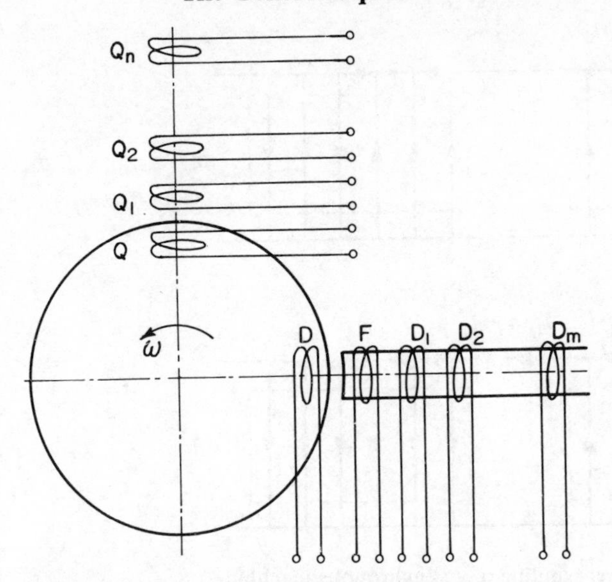

Fig. 4.3 Diagrams of a synchronous machine with many damper windings.

Consider a cage type damper winding of the type represented in Fig. 4.4 in which there are six bars in the pole face (shown dotted) connected by complete end rings. All the circuits are symmetrical with respect to both the direct and quadrature axes and in the general case can be represented in Fig. 4.3 by m direct-axis coils and n quadrature axis coils. However for practical work only an approximate representation, using a small number of damper coils, is feasible.

The winding in Fig. 4.4 requires three direct-axis coils and three quadrature-axis coils in the idealized machine. Fig. 4.4a shows the paths of the direct-axis currents and Fig. 4.4b those of quadrature-axis currents. In the practical machine the currents i_{d1}, i_{d2}, i_{d3}, i_{q1}, i_{q2}, i_{q3}, are superimposed in the bars and rings. Because of the symmetry about the axes there is no mutual action between the direct and quadrature-axis coils, but there is a mutual inductive coupling between the three coils of one axis. There is also a mutual resistance coupling because the current in any pair of bars flows through the end rings and introduces an ohmic drop in the circuit through any other pair of bars.

The eddy current paths in the iron can be considered as a cage winding having an infinite number of bars, but there is the

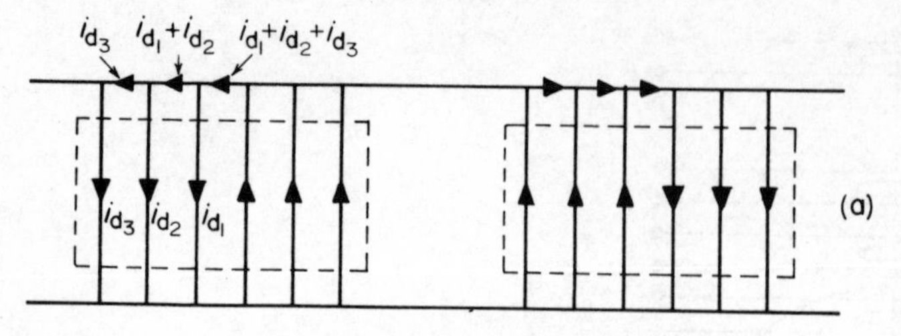

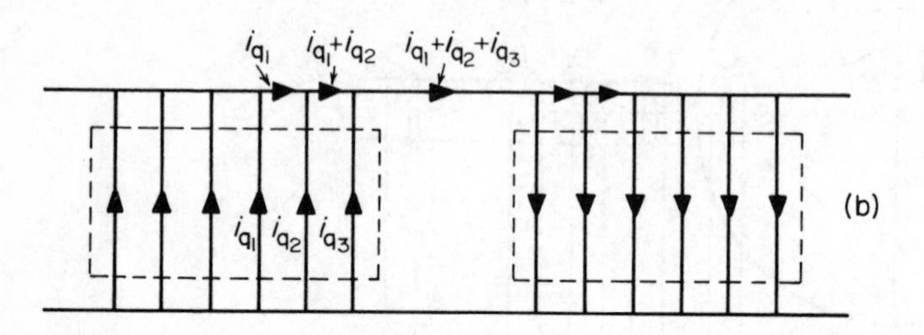

Fig. 4.4 Cage damper winding of a synchronous machine.

additional complication that the eddy currents cause the flux to be concentrated at the outer surface of the iron (the so-called skin effect) so that local saturation occurs. The problem of determining the effects of eddy currents is discussed in Chapter 10.

4.3.6 Direct-axis equations

The equations for a system of stationary mutually coupled coils comprising coils D, F and D_1 to D_m of Fig. 4.3 can be written down in terms of their self- and mutual inductances. For the rotating machine, however, equation (4.15) for the armature coil D differs from that for a stationary coil because of the rotational voltage term $\omega\psi_q$.

The flux linkage ψ_d depends only on the direct-axis currents and is given by:

$$\psi_d = L_{fd}i_f + L_{d1d}i_{d1} + L_{d2d}i_{d2} + \ldots L_d i_d \tag{4.17}$$

where L_{fd}, L_{d1d}, etc., are the mutual inductances between coils F, D_1 etc., and D; and L_d is the complete self-inductance of coil D.

The equations for the field and direct-axis damper coils are as follows:

$$\left.\begin{aligned}
u_f &= (R_f + L_{ff}p)i_f + L_{fd1}pi_{d1} + L_{fd2}pi_{d2} + \ldots L_{fd}pi_d \\
0 &= L_{fd1}pi_f + (R_{d1} + L_{d1}p)i_{d1} + (R_{d1d2} + L_{d1d2}p)i_{d2} \\
&\qquad + \ldots L_{d1d}pi_d \\
0 &= L_{fd2}pi_f + (R_{d1d2} + L_{d1d2}p)i_{d1} + (R_{d2} + L_{d2}p)i_{d2} \\
&\qquad + \ldots L_{d2d}pi_d
\end{aligned}\right\} \quad (4.18)$$

etc., where

R_f, R_{d1}, R_{d2}, L_{ff}, L_{d1}, L_{d2}, etc., are the resistance and self-inductances of the coils F, D1, D2, etc. L_{fd1} is the mutual inductance between coils F and D1, L_{d1d2} is the mutual inductance between coils D1 and D2, R_{d1d2} is the mutual resistance between coils D1 and D2, and similarly for other pairs of coils.

Eqns. (4.17) and (4.18) form a set of $(m + 2)$ equations containing $(m + 2)$ currents, as well as the quantities u_f and ψ_d. In many problems the currents in the field and damper windings are not required and may be eliminated by taking Laplace transforms (see Section 13.3). The relation between the Laplace transforms of ψ_d, i_d and u_f is of the form

$$\bar{\psi}_d = \frac{X_d(p)}{\omega_0}\bar{i}_d + \frac{G(p)}{\omega_0}\bar{u}_f \quad (4.19)$$

In the derivations of the following pages, the bars over the transformed variables are omitted, but it is understood that the algebraic manipulation of linear equations is based on the use of Laplace transforms.

The factor ω_0 in the denominator of Eqn. (4.19) is introduced in order that $X_d(p)$ shall have the dimensions of an impedance. $X_d(p)$ and $G(p)$ are functions of p, obtained in each case as the quotient of two determinants, each of which, when worked out, is a polynomial expression in p. $X_d(p)$, called the *direct-axis operational impedance*, is of the form:

$$X_d(p) = \frac{b_{(m+1)}p^{m+1} + b_m p^m + \ldots + b_0}{a_{(m+1)}p^{m+1} + a_m p^m + \ldots + a_0}$$

in which both numerator and denominator are of order $(m + 1)$. $G(p)$ has the same denominator as $X_d(p)$ but a different numerator of order m.

4.3.7 Quadrature-axis equations

The quadrature-axis equations are similar to those for the direct axis, but are simpler because there is no quadrature field winding. The symbols used correspond to those in the direct-axis equations:

$$\psi_q = L_{q1q} i_{q1} + L_{q2q} i_{q2} + \ldots L_q i_q \tag{4.20}$$

$$\left.\begin{aligned} 0 &= (R_{q1} + L_{q1}p) i_{q1} + (R_{q1q2} + L_{q1q2}p) i_{q2} + \ldots L_{q1q} p i_q \\ 0 &= (R_{q1q2} + L_{q1q2}p) i_{q1} + (R_{q2} + L_{q2}p) i_{q2} + \ldots L_{q2q} p i_q \end{aligned}\right\} \tag{4.21}$$

etc.

Elimination of the damper currents from the set of $(n + 1)$ equations gives:

$$\bar{\psi}_q = \frac{X_q(p)}{\omega_0} \bar{i}_q \tag{4.22}$$

where $X_q(p)$, the *quadrature-axis operational impedance,* is a function of p and is the quotient of two polynomials of order n.

4.4 Simplified equations of a synchronous machine with two damper coils

For most of the synchronous machine problems studied in the next few chapters the equations used are those of a simplified machine with one damper coil KD on the direct axis and one damper coil KQ on the quadrature axis, as in Fig. 4.2. Although it is only an approximation the results obtained are accurate enough for most purposes; for example, the concept of transient and subtransient reactances depends on this simplification. Even on this basis the working out of a practical solution is quite complicated enough, and it is often necessary to introduce still further approximations. On the other hand, the general equations of Section 4.3 provide the means of making a more detailed study if the discrepancies are too great; for example, in machines with complicated damping systems.

For the simplified machine, Eqns. (4.17) and (4.18) reduce to three direct-axis equations:

$$\left.\begin{aligned} \psi_d &= L_{fd} i_f + L_{dkd} i_{kd} + L_d i_d \\ u_f &= (R_f + L_{ff} p) i_f + L_{fkd} p i_{kd} + L_{fd} p i_d \\ 0 &= L_{fkd} p i_f + (R_{kd} + L_{kkd} p) i_{kd} + L_{dkd} p i_d \end{aligned}\right\} \quad (4.23)$$

Similarly Eqns. (4.20) and (4.21) reduce to two quadrature-axis equations:

$$\left.\begin{aligned} \psi_q &= L_{qkq} i_{kq} + L_q i_q \\ 0 &= (R_{kq} + L_{kkq} p) i_{kq} + L_{qkq} p i_q \end{aligned}\right\} \quad (4.24)$$

The functions $X_d(p)$, $G(p)$ and $X_q(p)$ of Eqns. (4.19) and (4.22) are much simpler than before and are of the form:

$$X_d(p) = \frac{b_2 p^2 + b_1 p + b_0}{a_2 p^2 + a_1 p + a_0}$$

$$G(p) = \frac{g_1 p + g_0}{a_2 p^2 + a_1 p + a_0}$$

$$X_q(p) = \frac{d_1 p + d_0}{c_1 p + c_0}$$

4.4.1 Equations using per-unit leakage inductances

Eqns. (4.17) to (4.24) hold whether the quantities are on a per-unit basis or not. They become easier to handle however, if per-unit quantities are used (see Section 1.2), and if, in addition, it is assumed that the three per-unit mutual inductances on the direct axis are all equal. In the rest of this book all equations are expressed in the per-unit system. The assumption about the mutual inductances has been accepted in the past as the basis of what may be called the conventional development of the theory, which has given good results for many purposes. It is however quite feasible to use more exact equations where necessary (see Section 4.7).

On a per-unit basis, with the above assumption:

$$L_{dkd} = L_{fd} = L_{fkd} = L_{md}$$

$$L_{qkq} = L_{mq}$$

where L_{md}, L_{mq} are the per-unit mutual inductances, or *magnetizing inductances*, on the two axes. The self-inductance of each coil is the sum of the mutual inductance and the leakage inductance (see Section 1.2). Hence:

$$\begin{aligned}
L_d &= L_{md} + L_a \\
L_{ff} &= L_{md} + L_f \\
L_{kkd} &= L_{md} + L_{kd} \\
L_q &= L_{mq} + L_a \\
L_{kkq} &= L_{mq} + L_{kq}
\end{aligned}$$

where L_a, L_f, L_{kd}, L_{kq} are per-unit leakage inductances. The armature leakage inductance L_a is the quantity determined on p. 71, and is assumed to have the same value for both axes.

Eqns. (4.23) and (4.24) now become:

$$\left.\begin{aligned}
\psi_d &= (L_{md} + L_a)i_d + L_{md}i_f + L_{md}i_{kd} \\
u_f &= L_{md}pi_d + [R_f + (L_{md} + L_f)]i_f + L_{md}pi_{kd} \\
0 &= L_{md}pi_d + L_{md}pi_f + [R_{kd} + (L_{md} + L_{kd})p]i_{kd}
\end{aligned}\right\} \quad (4.25)$$

$$\left.\begin{aligned}
\psi_q &= (L_{mq} + L_a)i_q + L_{mq}i_{kq} \\
0 &= L_{mq}pi_q + [R_{kq} + (L_{mq} + L_{kq})p]i_{kq}
\end{aligned}\right\} \quad (4.26)$$

Substitution of ψ_d and ψ_q in Eqns. (4.15) leads to the set of five Eqns. (4.27) for the synchronous machine. For the sake of generality, impressed voltages u_{kd} and u_{kq} are shown for the damper circuits, although in practice they are almost always zero.

An alternative form of the equations is used in Section 5.3. The field equation is expressed as

$$u_f = R_f i_f + p\psi_f$$

where

$$\psi_f = L_{md}i_d + (L_{md} + L_f)i_f + L_{md}i_{kd}$$

Flux linkages ψ_{kd} and ψ_{kq} are introduced in a similar manner (see Eqns. (5.13) to (5.15)).

$$\begin{bmatrix} u_{\mathrm{f}} \\ u_{\mathrm{kd}} \\ u_{\mathrm{d}} \\ u_{\mathrm{q}} \\ u_{\mathrm{kq}} \end{bmatrix} = \begin{bmatrix} R_{\mathrm{f}} + (L_{\mathrm{md}} + L_{\mathrm{f}})p & L_{\mathrm{md}}p & L_{\mathrm{md}}p & & \\ L_{\mathrm{md}}p & R_{\mathrm{kd}} + (L_{\mathrm{md}} + L_{\mathrm{kd}})p & L_{\mathrm{md}}p & & \\ L_{\mathrm{md}}p & L_{\mathrm{md}}p & R_{\mathrm{a}} + (L_{\mathrm{md}} + L_{\mathrm{a}})p & (L_{\mathrm{mq}} + L_{\mathrm{a}})\omega & L_{\mathrm{mq}}\omega \\ -L_{\mathrm{md}}\omega & -L_{\mathrm{md}}\omega & -(L_{\mathrm{md}} + L_{\mathrm{a}})\omega & R_{\mathrm{a}} + (L_{\mathrm{mq}} + L_{\mathrm{a}})p & L_{\mathrm{mq}}p \\ & & & L_{\mathrm{mq}}p & R_{\mathrm{kq}} + (L_{\mathrm{mq}} + L_{\mathrm{kq}})p \end{bmatrix} \begin{bmatrix} i_{\mathrm{f}} \\ i_{\mathrm{kd}} \\ i_{\mathrm{d}} \\ i_{\mathrm{q}} \\ i_{\mathrm{kq}} \end{bmatrix} \quad (4.27)$$

4.5 Equivalent circuits, operational impedances and frequency response loci

The equivalent circuits and operational impedances developed in this section apply only to the machine with one damper coil on each axis as shown in Fig. 4.2, but the same process can be applied to the more general machine of Fig. 4.4 [10].

Equivalent circuits

The principal complication in applying the equations of the previous section to practical problems, arises in handling the operational expressions $X_d(p)$, $X_q(p)$ and $G(p)$. For some purposes, especially steady a.c. problems, it is more convenient to derive these quantities from equivalent networks rather than by solving the equations.

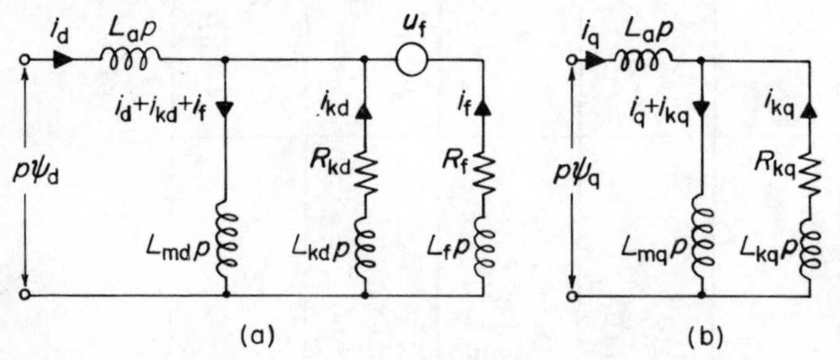

Fig. 4.5 Equivalent circuits of a synchronous machine
(a) Direct axis (b) Quadrature axis.

Considering first the direct axis, if Eqn. (4.17) is multiplied by p, $p\psi_d$ can be treated as an applied voltage. For a set of linear equations like Eqns. (4.17) and (4.18), in which all the mutual coefficients are equal in pairs, it is always possible to find a network to which the equations apply. $p\psi_d$ is the main applied voltage and u_f appears as an impressed voltage in a branch of the network which corresponds to the field circuit. Similarly another equivalent network can be found for the quadrature axis with $p\psi_q$ as the main applied voltage. It is evident that the quantities in the networks of Fig. 4.5 satisfy Eqns. (4.25) and (4.26).

The operational impedances $X_d(p)$, $X_q(p)$ and the function $G(p)$

For the simplified machine in Fig. 4.2, Eqn. (4.19) can be shown to be

$$\bar{\psi}_{\mathrm{d}} = \frac{1 + (T_4 + T_5)p + T_4 T_6 p^2}{1 + (T_1 + T_2)p + T_1 T_3 p^2} \cdot L_{\mathrm{d}} \bar{i}_{\mathrm{d}}$$

$$+ \frac{1 + T_{\mathrm{kd}} p}{1 + (T_1 + T_2)p + T_1 T_3 p^2} \cdot \frac{L_{\mathrm{md}} \bar{u}_{\mathrm{f}}}{R_{\mathrm{f}}} \qquad (4.28)$$

The values of the constants in Eqn. (4.28) are given below. They are expressed in terms of the reactances, each denoted by the symbol X with an appropriate suffix, and each equal to the corresponding inductance multiplied by ω_0, as listed on p. 88.

$$T_1 = \frac{1}{\omega_0 R_{\mathrm{f}}} (X_{\mathrm{md}} + X_{\mathrm{f}}),$$

$$T_2 = \frac{1}{\omega_0 R_{\mathrm{kd}}} (X_{\mathrm{md}} + X_{\mathrm{kd}}),$$

$$T_3 = \frac{1}{\omega_0 R_{\mathrm{kd}}} \left(X_{\mathrm{kd}} + \frac{X_{\mathrm{md}} X_{\mathrm{f}}}{X_{\mathrm{md}} + X_{\mathrm{f}}} \right),$$

$$T_4 = \frac{1}{\omega_0 R_{\mathrm{f}}} \left(X_{\mathrm{f}} + \frac{X_{\mathrm{md}} X_{\mathrm{a}}}{X_{\mathrm{md}} + X_{\mathrm{a}}} \right),$$

$$T_5 = \frac{1}{\omega_0 R_{\mathrm{kd}}} \left(X_{\mathrm{kd}} + \frac{X_{\mathrm{md}} X_{\mathrm{a}}}{X_{\mathrm{md}} + X_{\mathrm{a}}} \right),$$

$$T_6 = \frac{1}{\omega_0 R_{\mathrm{kd}}} \left(X_{\mathrm{kd}} + \frac{X_{\mathrm{md}} X_{\mathrm{f}} X_{\mathrm{a}}}{X_{\mathrm{md}} X_{\mathrm{a}} + X_{\mathrm{md}} X_{\mathrm{f}} + X_{\mathrm{a}} X_{\mathrm{f}}} \right),$$

$$T_{\mathrm{kd}} = \frac{X_{\mathrm{kd}}}{\omega_0 R_{\mathrm{kd}}},$$

The values of $X_{\mathrm{d}}(p)$ and $G(p)$ for the synchronous machine with one damper coil on each axis are found by comparing Eqns. (4.19) and (4.28) and equating coefficients. Expressions for $X_{\mathrm{d}}(p)$ and $G(p)$ can then be obtained in the form:

$$X_{\mathrm{d}}(p) = \frac{(1 + T_{\mathrm{d}}'p)(1 + T_{\mathrm{d}}''p)}{(1 + T_{\mathrm{d0}}'p)(1 + T_{\mathrm{d0}}''p)} \cdot X_{\mathrm{d}} \qquad (4.29)$$

$$G(p) = \frac{(1 + T_{\mathrm{kd}} p)}{(1 + T_{\mathrm{d0}}'p)(1 + T_{\mathrm{d0}}''p)} \cdot \frac{X_{\mathrm{md}}}{R_{\mathrm{f}}} \qquad (4.30)$$

where the new constants are determined by the identities:

$$(1 + T_{d0}'p)(1 + T_{d0}''p) \equiv 1 + (T_1 + T_2)p + T_1 T_3 p^2 \tag{4.31}$$

$$(1 + T_d'p)(1 + T_d''p) \equiv 1 + (T_4 + T_5)p + T_4 T_6 p^2 \tag{4.32}$$

The four new constants T_{d0}', T_{d0}'', T_d', T_d'' are the four principal time constants of the synchronous machine. Their values can be calculated accurately by solving two quadratic equations. More usually the values are calculated by making a further approximation, based on the fact that the per-unit resistance of the damper winding is generally much larger than that of the field winding. T_2 and T_3 are then much less than T_1, and the right-hand side of Eqn. (4.31) differs very little from $(1 + T_1 p)(1 + T_3 p)$. Hence T_{d0}' and T_{d0}'' are approximately equal to T_1 and T_3 respectively. Similarly T_d' and T_d'' are approximately equal to T_4 and T_6. The approximate values are given in the list on p. 89 together with the names of the four time constants.

The value of $X_q(p)$ is obtained by eliminating i_{kq} from Eqns. (4.26) and comparing the result with Eqn. (4.22):

$$X_q(p) = \frac{1 + T_q''p}{1 + T_{q0}''p} \cdot X_q \tag{4.33}$$

where T_{q0}'' and T_q'' have the values given on p. 89.

During steady-state operating conditions when all the variables in the two-axis reference frame are constant, $p = 0$ and according to Eqn. (4.29) the direct-axis operational impedance $X_d(p)$ equals the direct-axis synchronous reactance X_d. During rapid transients p tends to infinity and the limiting value of $X_d(p)$ is then defined as X_d'' and found from Eqn. (4.29) as

$$X_d'' = \frac{T_d' T_d''}{T_{d0}' T_{d0}''} \cdot X_d$$

In the absence of damper windings the value of $X_d(p)$ during rapid transients is defined as X_d' and found from Eqn. (4.29) as

$$X_d' = \frac{T_d'}{T_{d0}'} \cdot X_d$$

Similarly the quadrature-axis operational impedance $X_q(p)$ is equal to the quadrature-axis synchronous reactance X_q during steady-state conditions. During rapid transients, the limiting value of $X_q(p)$ is defined as X_q'' and found from Eqn. (4.33) as

$$X_q'' = \frac{T_q''}{T_{q0}''} \cdot X_q$$

In the absence of damper windings $X_q(p) = X_q$ for all values of p.

The direct-axis operational impedance is often used in the form of an operational admittance $Y_d(p)$ expanded into partial fractions. The admittance is found from Eqn. (4.29) as

$$Y_d(p) = \frac{1}{X_d(p)} = \frac{(1 + T_{d0}'p)(1 + T_{d0}''p)}{(1 + T_d'p)(1 + T_d''p)} \cdot Y_d \tag{4.34}$$

which, when expanded into partial fractions, becomes

$$Y_d(p) = \frac{1}{X_d}\left(1 + \frac{Ap}{1 + T_d'p} + \frac{Bp}{1 + T_d''p}\right) \tag{4.35}$$

where

$$A = -\frac{T_d'(1 - T_{d0}'/T_d')(1 - T_{d0}''/T_d')}{X_d(1 - T_d''/T_d')}$$

and

$$B = -\frac{T_d''(1 - T_{d0}'/T_d'')(1 - T_{d0}''/T_d''}{X_d(1 - T_d'/T_d'')}$$

Now T_d'', T_{d0}'' are small compared with T_d', T_{d0}' and the expression for $Y_d(p)$ in Eqn. (4.35) is approximately

$$Y_d(p) = \frac{1}{X_d} + \left(\frac{1}{X_d'} - \frac{1}{X_d}\right)\frac{T_d'p}{1 + T_d'p} + \left(\frac{1}{X_d''} - \frac{1}{X_d'}\right)\frac{T_d''p}{1 + T_d''p} \tag{4.36}$$

The partial fractions of $Y_d(p)$ can be considered to apply to a modified direct-axis equivalent circuit having three parallel branches only, the admittances of which correspond to the three terms of Eqn. (4.36). A similar modified quadrature-axis equivalent circuit exists for Eqn. (4.39).

Frequency response loci

For conditions relating to sinusoidal changes of frequency m, the operational admittance transforms to $Y_d(jm)$ which is found by replacing p by jm in Eqns. (4.34) to (4.36). Eqn. (4.36) then becomes

$$Y_d(jm) = \frac{1}{X_d} + \left(\frac{1}{X_d'} - \frac{1}{X_d}\right) \cdot \frac{jmT_d'}{1+jmT_d'} + \left(\frac{1}{X_d''} - \frac{1}{X_d'}\right) \cdot \frac{jmT_d''}{1+jmT_d''} \tag{4.37}$$

As m varies the locus of $Y_d(jm)$ in the complex plane, as given by Eqn. (4.37), is shown in Fig. 4.6a. It consists of a curve passing between two semi-circles which have their centres on the real axis.

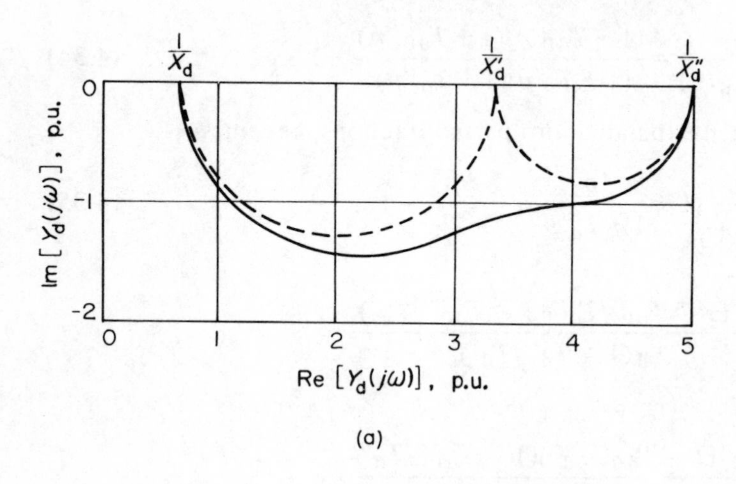

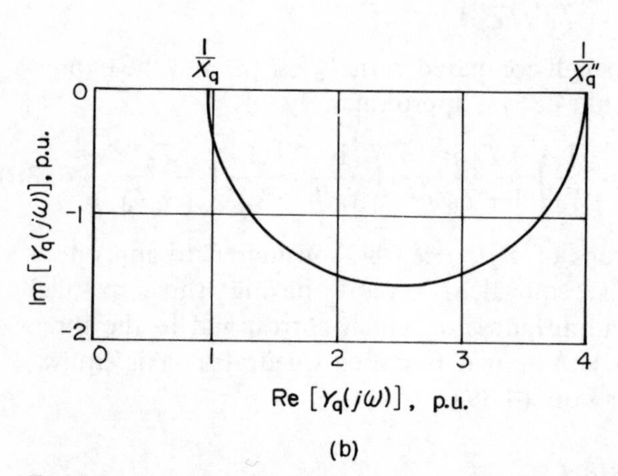

Fig. 4.6 Operational admittance frequency loci of a typical laminated-pole synchronous machine: (a) Direct axis (b) Quadrature axis.

The locus is analogous to the current locus diagram of a double-cage induction motor.

The quadrature-axis operational admittance $Y_q(p)$ is given by the reciprocal of Eqn. (4.33) as

$$Y_q(p) = \frac{1}{X_q(p)} = \frac{(1 + T_{q0}''p)}{(1 + T_q''p)} \cdot Y_q \tag{4.38}$$

which, when expanded into partial fractions, becomes

$$Y_q(p) = \frac{1}{X_q} + \left(\frac{1}{X_q''} - \frac{1}{X_q}\right) \cdot \frac{T_q''p}{1 + T_q''p} \tag{4.39}$$

The frequency response locus of $Y_q(jm)$, found by replacing p by jm in Eqn. (4.39), appears as a single semi-circle in Fig. 4.6b.

The term operational impedance, as applied to the quantities $X_d(p)$ and $X_q(p)$, is not a precise one. Strictly the term would apply more correctly to the quantities $pX_d(p)/\omega_0$ and $pX_q(p)/\omega_0$ which are the impedances of the equivalent circuits in Fig. 4.5. The same departure from strict terminology is retained in calling $Y_d(p)$ and $Y_q(p)$ 'operational admittances'.

4.6 Summary of the equations for the synchronous machine with two damper coils

For the study of practical synchronous machine problems it is often more convenient to retain the quantities ψ_d and ψ_q instead of using Eqns. (4.27)

$$\left.\begin{aligned} u_d &= p\psi_d + \omega\psi_q + R_a i_d \\ u_q &= -\omega\psi_d + p\psi_q + R_a i_q \end{aligned}\right\} \tag{4.15}$$

$$M_e = \frac{\omega_0}{2}(\psi_d i_q - \psi_q i_d) \tag{4.16}$$

$$\bar{\psi}_d = \frac{X_d(p)}{\omega_0} \cdot \bar{i}_d + \frac{G(p)}{\omega_0} \cdot \bar{u}_f \tag{4.19}$$

$$\bar{\psi}_q = \frac{X_q(p)}{\omega_0} \cdot \bar{i}_q \tag{4.22}$$

where

$$X_{\mathrm{d}}(p) = \frac{(1 + T_{\mathrm{d}}'p)(1 + T_{\mathrm{d}}''p)}{(1 + T_{\mathrm{d0}}'p)(1 + T_{\mathrm{d0}}''p)} \cdot X_{\mathrm{d}} \tag{4.29}$$

$$G(p) = \frac{(1 + T_{\mathrm{kd}}p)}{(1 + T_{\mathrm{d0}}'p)(1 + T_{\mathrm{d0}}''p)} \cdot \frac{X_{\mathrm{md}}}{R_{\mathrm{f}}} \tag{4.30}$$

$$X_{\mathrm{q}}(p) = \frac{(1 + T_{\mathrm{q}}''p)}{(1 + T_{\mathrm{q0}}''p)} \cdot X_{\mathrm{q}} \tag{4.33}$$

4.6.1 The constants of the synchronous machine

The following table contains a list of the fundamental constants and gives formulas for the constants in the above equations. All are per-unit values; ω_0 is the synchronous speed of the machine in electrical radians per second. Against each quantity is given the name by which it is known in the usually accepted terminology. The tests used for determining some of the constants experimentally are described in Ref. [39].

Fundamental machine constants

R_{a} = armature resistance.

R_{f} = field resistance.

R_{kd} = direct-axis damper resistance.

R_{kq} = quadrature-axis damper resistance.

$X_{\mathrm{md}} = \omega_0 L_{\mathrm{md}}$ = direct-axis magnetizing reactance.

$X_{\mathrm{mq}} = \omega_0 L_{\mathrm{mq}}$ = quadrature-axis magnetizing reactance.

$X_{\mathrm{a}} = \omega_0 L_{\mathrm{a}}$ = armature leakage reactance.

$X_{\mathrm{f}} = \omega_0 L_{\mathrm{f}}$ = field leakage reactance.

$X_{\mathrm{kd}} = \omega_0 L_{\mathrm{kd}}$ = direct-axis damper leakage reactance.

$X_{\mathrm{kq}} = \omega_0 L_{\mathrm{kq}}$ = quadrature-axis damper leakage reactance.

Time constants

$$T_{\mathrm{d0}}' = \frac{1}{\omega_0 R_{\mathrm{f}}}(X_{\mathrm{f}} + X_{\mathrm{md}})$$

= direct-axis transient open-circuit time constant.

$$T_d' = \frac{1}{\omega_0 R_f}\left(X_f + \frac{X_{md}X_a}{X_{md} + X_a}\right)$$

= direct-axis transient short-circuit time constant.

$$T_{d0}'' = \frac{1}{\omega_0 R_{kd}}\left(X_{kd} + \frac{X_{md}X_f}{X_{md} + X_f}\right)$$

= direct-axis subtransient open-circuit time constant.

$$T_d'' = \frac{1}{\omega_0 R_{kd}}\left(X_{kd} + \frac{X_{md}X_aX_f}{X_{md}X_a + X_{md}X_f + X_aX_f}\right)$$

= direct-axis subtransient short-circuit time constant.

$$T_{q0}'' = \frac{1}{\omega_0 R_{kq}}(X_{kq} + X_{mq})$$

= quadrature-axis subtransient open-circuit time constant.

$$T_q'' = \frac{1}{\omega_0 R_{kq}}\left(X_{kq} + \frac{X_{mq}X_a}{X_{mq} + X_a}\right)$$

= quadrature-axis subtransient short-circuit time constant.

$$T_{kd} = \frac{X_{kd}}{\omega_0 R_{kd}}$$

= direct-axis damper leakage time constant.

Derived reactances

$$X_d = X_a + X_{md}$$

= direct-axis synchronous reactance.

$$X_d' = X_d \cdot \frac{T_d'}{T_{d0}'} = X_a + \frac{X_{md}X_f}{X_{md} + X_f}$$

= direct-axis transient reactance.

$$X_d'' = X_d \cdot \frac{T_d'T_d''}{T_{d0}'T_{d0}''} = X_a + \frac{X_{md}X_fX_{kd}}{X_{md}X_f + X_{md}X_{kd} + X_fX_{kd}}$$

= direct-axis subtransient reactance.

$$X_q = X_a + X_{mq}$$

= quadrature-axis synchronous reactance.

$$X_q'' = X_q \frac{T_q''}{T_{q0}''} = X_a + \frac{X_{mq}X_{kq}}{X_{mq} + X_{kq}}$$

= quadrature-axis subtransient reactance.

4.6.2 Effect of series impedance

If an external impedance consisting of a resistance R_e and reactance X_e is in series with the machine, the machine equations can still be used for the combination. To do so, the external impedance is added to the internal impedance of the machine to yield a *modified machine* and R_a and X_a are replaced by R_t and X_t where

$$R_t = R_a + R_e$$
$$X_t = X_a + X_e$$

All the reactances and time constants are modified by this change. Each of the derived reactances is increased by X_e. Also since T_{d0}' and T_{d0}'' do not depend on X_e, T_d' and T_d'' can be calculated from the above formulas for X_d' and X_d'', if X_d, X_d' and X_d'' are replaced by $(X_d + X_e)$, $(X_d' + X_e)$ and $(X_d'' + X_e)$. T_q'' is obtained in a similar manner.

4.7 Modified equations with more accurate coupling between field and damper windings

The equations derived in Section 4.4 are based on the assumption that the three direct-axis mutual inductances between armature, field and damper windings are all equal. However, Fig. 8.4 shows that, although this theory gives good results in calculating the armature currents after a sudden short-circuit, there is considerable error in determining the field current. Canay [42] has shown that a more accurate value is obtained if an additional parameter is introduced in the equations to allow the mutual inductance between field and damper to differ from the other two. The mutual inductances between armature and field and between armature and damper would still be assumed equal. In Fig. 4.7, which indicates the various mutual and leakage fluxes, corresponding to the inductances, L_{kf} corresponds to the leakage flux which

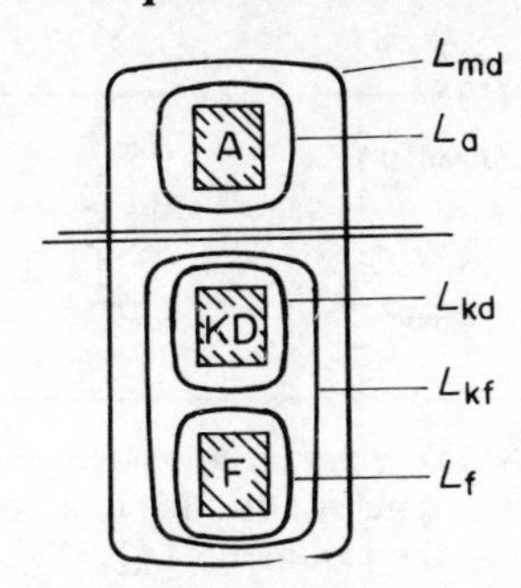

Fig. 4.7 Diagram showing the various mutual and leakage fluxes of a synchronous machine.

links F and KD, but not D. The modified equations are obtained from Eqn. (4.25) by adding L_{kf} where appropriate

$$\left.\begin{aligned}\psi_d &= (L_{md} + L_a)i_d + L_{md}i_f + L_{md}i_{kd}\\ u_f &= L_{md}pi_d + [R_f + (L_{md} + L_f + L_{kf})p]i_f\\ &\quad + (L_{md} + L_{kf})pi_{kd}\\ 0 &= L_{md}pi_d + (L_{md} + L_{kf})pi_f + [R_{kd}\\ &\quad + (L_{md} + L_{kd} + L_{kf})p]i_{kd}\end{aligned}\right\} \quad (4.40)$$

Strictly a system of three coupled coils requires three self-inductances, three mutual inductances and three resistances to define it completely. Since, when expressed in per-unit, the six inductances differ from each other by small amounts only, it is preferable to select the smallest mutual inductance L_{md} (armature to field) and to express each of the others as the sum of L_{md} and a leakage inductance. As stated above the mutual inductance between armature and damper is taken to be L_{md}. The equations now contain five independent inductances, compared with four in Eqn. 4.25. Fig. 4.8 shows the modified equivalent circuit (Fig. 4.5a with L_{kf} added), which is similar to that of a double-cage induction motor (Fig. 10.13).

Following the method of Section 4.5 and making the same approximations, modified formulas for the derived reactances and time constants can be calculated.

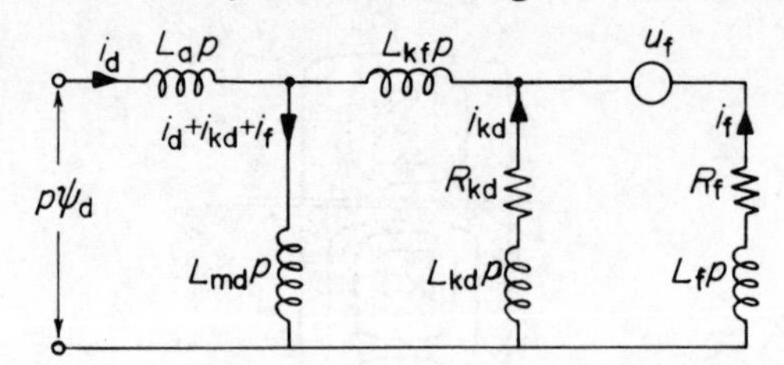

Fig. 4.8 Direct-axis equivalent circuit of a synchronous machine including the parameter L_{kf}

$$\left.\begin{aligned}
X_d &= X_{md} + X_a \\
X_d' &= X_a + \frac{X_{md}(X_f + X_{kf})}{X_{md} + X_f + X_{kf}} \\
X_d'' &= X_a + \frac{X_{kd}X_fX_{md} + X_{kd}X_{kf}X_{md} + X_{md}X_fX_{kf}}{X_{md}X_f + X_{md}X_{kd} + X_{kd}X_f + X_{kd}X_{kf} + X_fX_{kf}}
\end{aligned}\right\} \quad (4.41)$$

$$\left.\begin{aligned}
T_d' &= \frac{1}{\omega_0 R_f}\left(X_f + X_{kf} + \frac{X_{md}X_a}{X_{md} + X_a}\right) \\
T_d'' &= \frac{1}{\omega_0 R_{kd}} \\
&\quad \left(X_{kd} + \frac{X_{md}X_{kf}X_f + X_aX_fX_{md} + X_aX_fX_{kf}}{X_fX_{md} + X_fX_a + X_{kf}X_{md} + X_{kf}X_a + X_{md}X_a}\right) \\
T_{d0}' &= \frac{1}{\omega_0 R_f}(X_{md} + X_f + X_{kf}) \\
T_{d0}'' &= \frac{1}{\omega_0 R_{kd}}\left(X_{kd} + \frac{X_f(X_{md} + X_{kf})}{X_{kf} + X_f + X_{md}}\right)
\end{aligned}\right\} \quad (4.42)$$

The first five of the above parameters can also be determined from a measured short-circuit oscillogram, as explained in Section 8.2. These values, together with the quadrature-axis parameters, obtained by calculation from the design or by measurement, suffice for most of the methods of analysis in Chapters 7 and 8, which make use of the operational impedance functions. However, the more fundamental Eqns. 4.25 and 4.26, and the state-space formulation derived from them in Section 5.3, include the resistance and the fundamental mutual and leakage inductances. If it is desired to deduce these fundamental values from the measured reactances and time constants, it is necessary to solve Eqns. (4.41) and (4.42) for them.

As stated on p. 26 it is generally recognized that there is no satisfactory method of measuring the armature leakage inductance L_a and a calculated or estimated value is taken. In the usual method, with X_{kf} neglected, the three direct-axis Eqns. (4.41) are sufficient to determine X_{md}, X_f and X_{kd}, but for the new method further information is needed to determine X_{kf}. The resistances are determined from Eqn. (4.42).

It is a curious fact that, in making a short-circuit test, it has been the practice for many years to record, in addition to the armature current oscillograms, an oscillogram of the field current, of which little or no use is ever made. This oscillogram provides the necessary information for determining X_{kf}. Referring to Eqn. (8.16), the coefficient of $\epsilon^{-t/T_d'}$ provides no new information but the coefficient of $\epsilon^{-t/T_d''}$ determines a new quantity K.

$$K = \left(1 - \frac{T_{kd}}{T_d''}\right) = \frac{X_{kd}}{\omega_0 R_{kd} T_d''} \tag{4.43}$$

The four Eqns. (4.41) and (4.43) are now sufficient to determine the four reactances required [66].

If the mutual inductances from armature to field and damper were not assumed equal, a further piece of information would be needed, and could be obtained from an oscillogram of damper current. Such a measurement however is not possible in practice. Moreover, the assumption is a reasonable one, particularly as the error would be mainly in the damper current. The field current is more important than the damper current because it affects the external regulator circuit.

A further precaution may be necessary. The above procedure would make it possible to obtain correct parameters, either the transient reactances and time constants or the resistances and the mutual and leakage inductances, for use with the new method, but the values would be different from those given by established design methods. Since the design methods for calculating the resistances and the leakage inductances have generally been built up from past experience using empirical factors to obtain agreement with the test results, a review of design methods for use with the new theory may well be needed.

4.8 General equations of the induction motor

The induction motor has a uniform air gap and both of its windings are a.c. windings distributed in slots. The great majority of practical induction motors have the primary winding on the stator and the secondary winding on the rotor. The reverse arrangement is however quite possible. As explained on p. 5, the theory depends only on relative motion, and all the diagrams of primitive machines are drawn with the reference axis stationary.

Because of the uniform air gap, two alternative theoretical treatments are possible, according to whether the axes of the reference frame are assumed to be attached to the secondary or the primary member: If the reference frame of the idealized machine is attached to the secondary member, the induction motor can be considered as a special case of the synchronous machine, and the equations are those of Eqns. (4.27) in a simplified form. If, on the other hand, the reference frame is attached to the primary member, different equations are obtained and different transformations are needed. The equations in the latter form are referred to as Kron's equations [8].

4.8.1 Park's equations for the induction motor

The primitive diagram for an induction motor with the primary winding rotating is shown in Fig. 4.9, in which the suffixes 1 and 2 are used to denote primary and secondary. Because of the uniform air gap, corresponding inductances are the same on both axes. There are no self-evident direct and quadrature axes but the analysis is simplified by choosing the d-axis to coincide with the axis of phase A_2 on the secondary winding. The current transformation from frame (a_2,b_2,c_2) to (d_2,q_2,z_2) is Eqn. (4.10).

The three-phase primary coils A_1, B_1 and C_1 are replaced by fictitious two-axis coils D_1 and Q_1 and the transformation equations are those given in Eqns. (4.4) to (4.7) which include the zero sequence component. The voltage and torque equations, if zero sequence equations are omitted, are

$$\begin{bmatrix} u_{d2} \\ u_{d1} \\ u_{q1} \\ u_{q2} \end{bmatrix} = \begin{bmatrix} R_{d2} + L_{22}p & L_m p & & \\ L_m p & R_{d1} + L_{11}p & L_{11}\omega & L_m \omega \\ -L_m \omega & -L_{11}\omega & R_{q1} + L_{11}p & L_m p \\ & & L_m p & R_{q2} + L_{22}p \end{bmatrix} \begin{bmatrix} i_{d2} \\ i_{d1} \\ i_{q1} \\ i_{q2} \end{bmatrix} \quad (4.44)$$

$$M_e = \frac{\omega_0}{2}(L_m i_{d2} i_{q1} - L_m i_{d1} i_{q2}) \quad (4.45)$$

where L_m is the magnetizing inductance and L_{11} and L_{22} are the self-inductances.

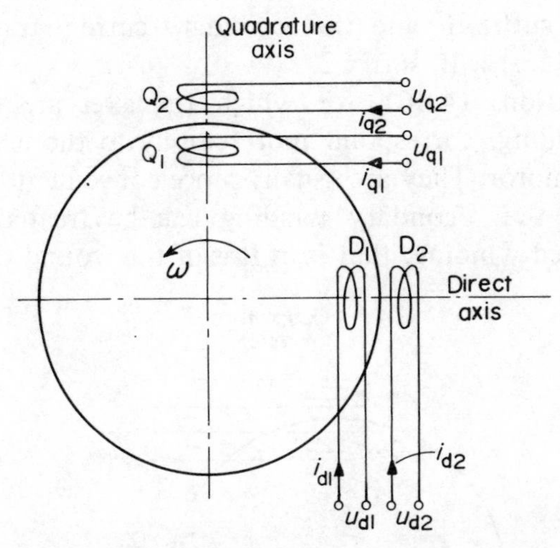

Fig. 4.9 Diagram of an induction motor with the reference axes attached to the secondary member.

4.8.2 Kron's equations for the induction motor

The alternative arrangement in which the reference frame is attached to the primary member with the direct axis chosen to coincide with the primary phase A_1, is shown in Fig. 4.10. The only essential difference compared with Fig. 4.9 is that the primary and secondary applied voltages are changed over. The form of the equations is therefore exactly as before with the suffixes interchanged. The voltage and torque equations are:

$$\begin{bmatrix} u_{d1} \\ u_{d2} \\ u_{q2} \\ u_{q1} \end{bmatrix} = \begin{bmatrix} R_{d1}+L_{11}p & L_m p & & \\ L_m p & R_{d2}+L_{22}p & L_{22}\omega & L_m\omega \\ -L_m\omega & -L_{22}\omega & R_{q2}+L_{22}p & L_m p \\ & & L_m p & R_{q1}+L_{11}p \end{bmatrix} \begin{bmatrix} i_{d1} \\ i_{d2} \\ i_{q2} \\ i_{q1} \end{bmatrix} \quad (4.46)$$

$$M_e = \frac{\omega_0}{2}(L_m i_{d1} i_{q2} - L_m i_{d2} i_{q1}) \quad (4.47)$$

The primary current transformations are obtained from Eqns. (4.10) with suffix 1, and the secondary current transformations from Eqns. (4.4) with suffix 2.

The equations (4.46), for which the axes are fixed to the primary winding, correspond more closely to the usual construction of the motor. They are usually more convenient for analytical purposes if the secondary winding can be treated as a single short-circuited winding, that is, if it is of the wound type with slip

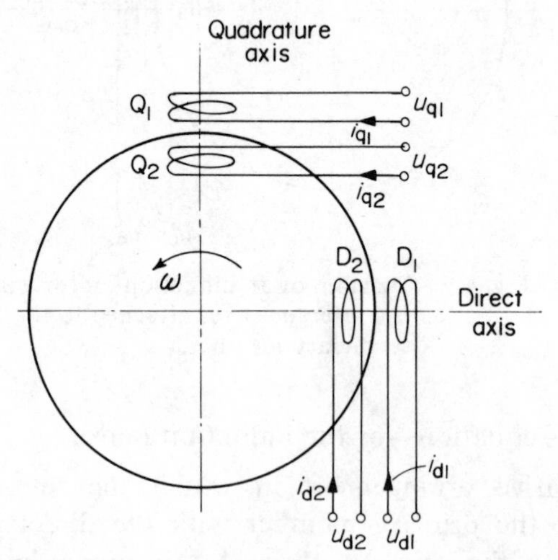

Fig. 4.10 Diagram of an induction motor with the reference axes attached to the primary member.

rings, or has a single cage winding. If there is appreciable deep-bar effect in a cage winding or if an external injected voltage is applied to a slip-ring winding, the form of Eqn. (4.44) is likely to be more convenient.

4.8.3 Cage windings

Any possible arrangement of cage bars and end-rings, including double-cages, can be dealt with by a method similar to that explained on p. 75. However, any cage winding with m equally spaced bars forming an m-phase winding can be treated much more simply by transforming it into an equivalent two-phase winding. For a rotating winding in the primitive machine the direct and quadrature-axis voltages are:

$$u_{\mathrm{d}} = \frac{2}{m}\left\{u_{\mathrm{a}}\cos\theta + u_{\mathrm{b}}\cos\left(\theta - \frac{2\pi}{m}\right) + \ldots + u_{\mathrm{m}}\cos\left(\theta - \frac{2(m-1)\pi}{m}\right)\right\}$$

$$u_{\mathrm{q}} = \frac{2}{m}\left\{u_{\mathrm{a}}\sin\theta + u_{\mathrm{b}}\sin\left(\theta - \frac{2\pi}{m}\right) + \ldots + u_{\mathrm{m}}\sin\left(\theta - \frac{2(m-1)\pi}{m}\right)\right\} \tag{4.48}$$

For a stationary winding, the values can be obtained from Eqns. (4.48) by putting $\theta = 0$.

In order to define the set of variables completely it would be necessary to specify additional voltages and currents, $(m - 2)$ in number, corresponding to the zero-sequence quantities u_{z} and i_{z} in the three-phase case. These components are similar to the quantities obtained when resolving an m-phase system into symmetrical components, and are determined by independent equations. The practical cage winding consists of many coupled circuits interconnected through the end-rings, each circuit being strictly a mesh in a network and each current being the current in a bar. Since the impressed voltage in every circuit is zero, all the symmetrical component voltages are also zero. Consequently all the current components except i_{d} and i_{q} are also zero. Hence the cage is equivalent to the two-phase winding formed by coils D_2 and Q_2 in Fig. 4.10 and for most purposes it is not necessary to introduce the actual currents at all.

The equations of a double-cage induction motor can be written down by including an additional pair of two-phase coils in Fig. 4.9 or 4.10 and adding two more voltage equations (see Section 11.2).

Chapter Five
Types of Problem and Methods of Solution and Computation

5.1 Classification of problems and methods of solution

For a machine represented in the primitive diagram by n coils there are n voltage equations and a torque Eqn. (1.6). If the n applied voltages and the applied torque are known, as well as initial conditions, the $(n + 1)$ equations are sufficient to determine the n currents and the speed. Hence theoretically the performance of the machine is completely determined.

In the general case, the equations, which contain product terms involving the speed and the currents, are non-linear differential equations and have to be solved by numerical integration. For particular conditions, however, considerable simplifications can often be made, and the types of problem encountered in practice can be classified in the manner described in the following paragraphs and tabulated in Table 5.1. Some of the mathematical methods mentioned in Table 5.1 are discussed in Chapter 13.

Under steady conditions the speed is constant and the voltage equations can be dealt with independently of the torque equations. The voltage equations reduce under steady conditions to a set of n ordinary linear algebraic equations, containing real variables for d.c. conditions (item 1), or complex variables for a.c. conditions (item 2).

Under transient conditions, when voltages and currents may vary in any manner, the problem is greatly simplified if the speed

Table 5.1

Condition	*Equation*	*Method of solution*	*Solution*
SPEED CONSTANT. VOLTAGE EQUATIONS ONLY			
1. Steady d.c.	Real algebraic equations	Real algebra	Real numbers
2. Steady a.c.	Complex algebraic equations	Complex algebra Phasor diagram	Complex numbers
3. Transient	Linear D.E.'s Constant coefficients	Laplace transforms Eigen values and Eigen vectors	Functions of t
SPEED A KNOWN FUNCTION OF t. VOLTAGE EQUATIONS ONLY			
4. Transient	Linear D.E.'s Variable coefficients	Step-by-step calculation	Functions of t.
SPEED UNKNOWN. VOLTAGE AND TORQUE EQUATIONS			
5. Transient	Non-linear D.E.'s	As 4	Functions of t
6. Small changes	Linear D.E.'s Constant coefficients	As 3	Functions of t
7. Small oscillations	Complex algebraic equations	As 2	Complex numbers

has a known constant value (item 3), because the voltage equations, which can then still be handled independently of the torque equation, are linear differential equations with constant coefficients.

When the voltages are known the equations can be solved, either by algebraic methods for steady-state problems, or by operational methods for transient problems, and the currents thus obtained can be substituted in the torque equation. If the speed is constant the externally applied torque is obtained directly because it is equal to the electrical torque.

The next stage in difficulty arises if the speed varies but is a known function of time (item 4). It is then still possible to solve the voltage equations separately, but the coefficients, which depend on the speed, are not constant. The equations are linear equations with variable coefficients for which the simple operational method does not apply. Usually only numerical solutions are possible. Once the currents are determined the electrical torque is found by substitution in Eqn. (1.10). If the applied torque is required, the inertia term in Eqn. (1.6) must be added.

The most difficult problems are those for which the speed is an unknown variable (item 5). This condition covers the most general type of transient problem, for which all the $(n + 1)$ equations must be handled together by a numerical method.

The rather special conditions covered by items 6 and 7 can be treated in a much simpler manner than the general case. For a small variation relative to a given steady condition, the non-linear differential equations, including both voltage and torque equations, can, to a close approximation, be converted into linear ones. When the variations take the form of sinusoidal oscillations at a known frequency, they can be further reduced to complex algebraic equations. The method of deriving the equations for small changes or small oscillations is explained in detail in Section 2.3.

The equations are frequently used to study the stability of a system. Transient stability, which depends on the response to a large disturbance, is studied by using the full non-linear equations (item 5). For steady-state stability, which depends on the response to a small disturbance, the equations for small changes or small oscillations are used (items 6 and 7).

For any given problem there is a choice between a fully accurate method using the complete equations and various simplified approximate methods in which some of the terms in the equations are neglected.

5.2 Modified machine equations in terms of rotor angle δ

When a synchronous machine operates in conjunction with an external supply, the definition of δ, given on p. 53 applies, but under transient conditions δ is a variable function of time. If the machine is not connected directly to an infinite bus, there may be one at another point in the system or there may only be a fictitious one, assumed for calculation purposes.

The position θ_r of the reference axis at any instant is

$$\theta_r = \omega_0 t$$

and the rotor position, speed, slip and acceleration are respectively,

$$\left.\begin{aligned} \theta &= \omega_0 t - \delta \\ \omega &= \omega_0 - p\delta \\ s &= p\delta/\omega_0 \\ p^2\theta &= -p^2\delta. \end{aligned}\right\} \tag{5.1}$$

The torque equation is

$$\left.\begin{aligned} M_t &= M_e - \frac{2H}{\omega_0}\cdot p^2\delta \\ &= \frac{\omega_0}{2}(\psi_d i_q - \psi_q i_d) - \frac{2H}{\omega_0}p^2\delta. \end{aligned}\right\} \tag{5.2}$$

The armature voltage equations are

$$\left.\begin{aligned} u_d &= p\psi_d + \omega_0\psi_q + R_a i_d - \psi_q p\delta \\ u_q &= -\omega_0\psi_d + p\psi_q + R_a i_q + \psi_d p\delta. \end{aligned}\right\} \tag{5.3}$$

Equations (5.3) are general for any balanced supply conditions. If the machine is connected directly to the infinite bus, the axis voltages are determined as follows; The voltage in phase A of a balanced three-phase supply of frequency $\frac{\omega_0}{2\pi}$ is

$$\begin{aligned} u_a &= U_m \sin \omega_0 t \\ &= U_m \sin\delta \cos\theta + U_m \cos\delta \sin\theta \end{aligned}$$

The transformation equation (4.7) relating the voltage in phase A to the axis voltage is

$$u_a = u_d \cos\theta + u_q \sin\theta$$

The two values of u_a must be identical for all θ. Hence

$$\left.\begin{aligned} u_d &= U_m \sin\delta \\ u_q &= U_m \cos\delta. \end{aligned}\right\} \tag{5.4}$$

Substituting in Eqn. (5.3)

$$\left.\begin{aligned} U_m \sin\delta &= p\psi_d + \omega_0\psi_q + R_a i_d - \psi_q p\delta \\ U_m \cos\delta &= -\omega_0\psi_d + p\psi_q + R_a i_q + \psi_d p\delta. \end{aligned}\right\} \tag{5.5}$$

If only the conditions in the generator itself are considered, the angle δ_t between the rotor position and the angular position determined by the terminal voltage is often called the 'generator load angle'. However, for the general transient condition, the terminal angle is itself variable relative to the angle of the infinite bus, (real or fictitious). For calculation purposes it is essential to define δ in the above equations as the angle relative to the infinite bus, since this is the mechanical angle which determines the inertia torque.

5.3 The state variable method and the state-space concept

The *state variable* theory and its associated techniques are of great value in the study of complicated control systems. A multi-machine power system, in which each generator has a voltage regulator and a speed governor, provides a good example of such a system. A brief introduction to this theory, including the construction of a state-space model of the synchronous machine, is given in the following pages. For a fuller treatment, some of the many excellent texts [35, 45] available should be consulted.

The *state variable equations* of a system are defined as a set of n first order differential equations, having the minimum number of state variables $x_1(t) \ldots x_n(t)$ required to define the system. In order to calculate the variables for a given condition of operation, the state equations are used in conjunction with the *control variables* (excitation or torque) applied externally, as well as the relevant initial conditions. The m control variables (often alternatively called *input variables*) are $z_1(t) \ldots z_m(t)$. It may also be necessary to derive separate *output variables* $y_1(t) \ldots y_r(t)$.
In matrix form

$$\boldsymbol{x} = \begin{bmatrix} x_1(t) \\ x_2(t) \\ \vdots \\ x_n(t) \end{bmatrix}$$

and similarly for $\boldsymbol{y}$ and $\boldsymbol{z}$.

The equations constitute an algebraic statement of the conditions in the system. An alternative geometrical statement is provided by the *state-space* concept. The state variables are considered to be the components of an n-dimensional vector in an n-dimensional space. The curve followed by the end point of the vector as time passes is called a *trajectory*. It is not easy to visualize a space of more than three dimensions, but the concept is a useful one, because it can be illustrated by diagrams in two or three dimensions. In any mechanical system, for which one of the equations is an equation of motion giving the acceleration $\ddot{x}$, the displacement x and the velocity $\dot{x}$ are two of the important state variables. Even when there are many other state variables a good deal of information about the behaviour of a system is provided by a two-dimensional state-space diagram relating x and $\dot{x}$ (often referred to as a *phase-plane diagram*). The concept is particularly useful for a system including synchronous machines, for which the phase-plane diagram relating the angular displacement and velocity, δ and $\dot{\delta}$, is very useful in the assessment of stability.

Each of the n differential equations gives the time derivative of each state variable as a function of the state and control variables. In the general non-linear case

$$\dot{\boldsymbol{x}}(t) = \boldsymbol{F}(\boldsymbol{x},\boldsymbol{z}) \tag{5.6}$$

$$\boldsymbol{y}(t) = \boldsymbol{C}(\boldsymbol{x},t) \tag{5.7}$$

For a linear system

$$\dot{\boldsymbol{x}} = \boldsymbol{A}\boldsymbol{x} + \boldsymbol{B}\boldsymbol{z} \tag{5.8}$$

$$\boldsymbol{y} = \boldsymbol{C}\boldsymbol{x} \tag{5.9}$$

The term *linear* is here used in the physical sense that all effects are linearly proportional to the quantities causing them, that is, that the equations are mathematically linear with constant coefficients. A system for which the equations are mathematically linear with variable coefficients is regarded as physically non-linear.

In Eqns. (5.6) to (5.9) $\boldsymbol{F}(\boldsymbol{x},\boldsymbol{z})$ is a set of n functions and $\boldsymbol{C}(\boldsymbol{x},t)$ is a set of r functions. $\boldsymbol{A}$, $\boldsymbol{B}$, $\boldsymbol{C}$ are constant matrices as follows.

$\boldsymbol{A}$ = $n \times n$ *system matrix*

$\boldsymbol{B}$ = $n \times m$ *control matrix*

$\boldsymbol{C}$ = $r \times n$ *output matrix.*

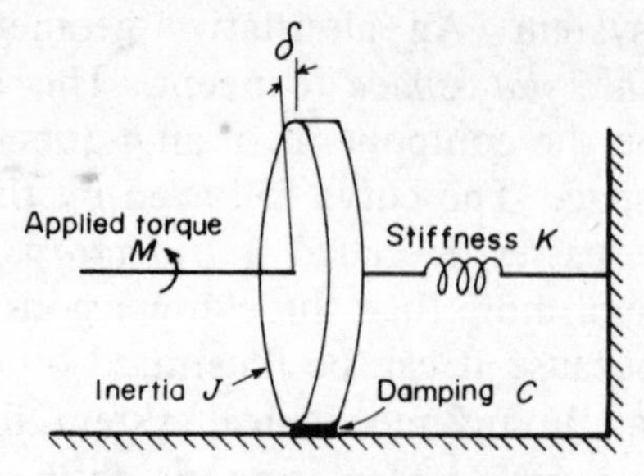

Fig. 5.1 Spring-mass-damper system.

Consider the simple oscillating mechanical system shown in Fig. 5.1. The equation of motion, assuming zero initial conditions, is

$$J\ddot{\delta} + C\dot{\delta} + K\delta = M(t)$$

$$\text{or} \quad \ddot{\delta} + a\dot{\delta} + b^2\delta = \frac{M(t)}{J}$$

If δ and $\dot{\delta}$ are chosen as the state variables x_1 and x_2 respectively, and $M(t)$ as the input variable z, the equations in state space form are

$$\dot{x} = \begin{bmatrix} 0 & 1 \\ -b^2 & -a \end{bmatrix} x + \begin{bmatrix} 0 \\ \dfrac{1}{J} \end{bmatrix} z \tag{5.10}$$

The choice of states used, which may be made in many ways, depends on the type of system and consideration should be given to the concepts of *controllability* and *observability*. These concepts basically are concerned with the ability to affect (controllable) and detect (observable) the dynamics of the system from the inputs and outputs. While most physical systems are controllable and observable the state-space models do not necessarily have these properties.

5.3.1 State-space model of a synchronous machine

The equations of a synchronous machine are readily put into state variable form. The state variables can be selected in many possible ways, but the choice of the flux linkage functions, as in Eqns. (5.11), leads directly to a derivation of the state-space equations, because the equation for each coil contains the derivative of the flux linking it, and no other.

$$\left.\begin{array}{lll} x_1 = \delta & x_3 = \omega_0 \psi_d & x_6 = \omega_0 \psi_q \\ x_2 = \dot{\delta} & x_4 = \omega_0 \psi_f & x_7 = \omega_0 \psi_{kq} \\ & x_5 = \omega_0 \psi_{kd} & \end{array}\right\} \tag{5.11}$$

$$\left.\begin{array}{l} z_1 = M_t = \text{shaft torque} \\ z_2 = u_f = \text{field voltage.} \end{array}\right\} \tag{5.12}$$

Eqns. (5.3), (4.25) and (4.26) are rearranged in Eqns. (5.13) to (5.15) by introducing the additional variables ψ_f, ψ_{kd} and ψ_{kq} as explained on p. 80.

$$\left.\begin{array}{l} \dot{\psi}_d = u_d - \omega_0 \psi_q - R_a i_d + \psi_q \cdot \dot{\delta} \\ \dot{\psi}_q = u_q + \omega_0 \psi_d - R_a i_q - \psi_d \cdot \dot{\delta} \\ \dot{\psi}_f = u_f - R_f i_f \\ \dot{\psi}_{kd} = -R_{kd} i_{kd} \\ \dot{\psi}_{kq} = -R_{kq} i_{kq}. \end{array}\right\} \tag{5.13}$$

The currents are related to the fluxes by two sets of linear equations.

$$\begin{bmatrix} \omega_0 \psi_d \\ \omega_0 \psi_f \\ \omega_0 \psi_{kd} \end{bmatrix} = \begin{bmatrix} X_{md} + X_a & X_{md} & X_{md} \\ X_{md} & X_{md} + X_f & X_{md} \\ X_{md} & X_{md} & X_{md} + X_{kd} \end{bmatrix} \begin{bmatrix} i_d \\ i_f \\ i_{kd} \end{bmatrix} \tag{5.14}$$

$$\begin{bmatrix} \omega_0 \psi_q \\ \omega_0 \psi_{kq} \end{bmatrix} = \begin{bmatrix} X_{mq} + X_a & X_{mq} \\ X_{mq} & X_{mq} + X_{kq} \end{bmatrix} \begin{bmatrix} i_q \\ i_{kq} \end{bmatrix} \tag{5.15}$$

Inversion of the matrices gives the currents in terms of the fluxes.

$$\begin{bmatrix} i_d \\ i_f \\ i_{kd} \end{bmatrix} = \omega_0 \begin{bmatrix} Y_{1d} & Y_{4d} & Y_{5d} \\ Y_{4d} & Y_{2d} & Y_{6d} \\ Y_{5d} & Y_{6d} & Y_{3d} \end{bmatrix} \begin{bmatrix} \psi_d \\ \psi_f \\ \psi_{kd} \end{bmatrix} \tag{5.16}$$

$$\begin{bmatrix} i_q \\ i_{kq} \end{bmatrix} = \omega_0 \begin{bmatrix} Y_{1q} & Y_{3q} \\ Y_{3q} & Y_{2q} \end{bmatrix} \begin{bmatrix} \psi_q \\ \psi_{kq} \end{bmatrix} \tag{5.17}$$

The quantities all have a physical meaning; in particular, the Y coefficients are self and mutual admittances. Y_{1d} and Y_{1q} are reciprocals of the subtransient reactances X_d'' and X_q''.

The torque M_t is given by

$$M_t = \frac{\omega_0}{2}(\psi_d i_q - \psi_q i_d) - J\ddot{\delta} - K\dot{\delta} \tag{5.18}$$

where K is a constant damping factor.

Substitution of the currents of equations (5.16) and (5.17) in (5.13) and (5.18) leads to the state variable equation (5.19), where

$$M_e = \tfrac{1}{2}\{Y_{1q}x_3x_6 + Y_{3q}x_3x_7 - Y_{1d}x_6x_3 - Y_{4d}x_6x_4 - Y_{5d}x_6x_5\} \tag{5.20}$$

Equation (5.19) can also be written as

$$\dot{\boldsymbol{x}} = \boldsymbol{A}\boldsymbol{x} + \boldsymbol{F}(\boldsymbol{x}) + \boldsymbol{B}\boldsymbol{z} \tag{5.21}$$

It can be seen that the great majority of the terms are linear, but there is a small number of non-linear terms which are separated out in the part indicated by $\boldsymbol{F}(\boldsymbol{x})$.

The terminal voltage components u_d and u_q depend on the external network. For a multi-machine system they must be transformed to a network reference frame and combined with the voltage components of the network and the other machines, as explained in Section 9.5. If the machine is connected to an infinite bus through an external impedance, the impedance can be added to that of the generator to form a modified machine for which the terminal voltage components are given by Eqn. (5.4).

If the input variables depend on voltage or speed regulators for which the equations are known, the regulator and machine equations can be combined together in single larger matrices $\boldsymbol{A}$ and $\boldsymbol{F}(\boldsymbol{x})$ for which the new input variables are the inputs to the regulators.

$$
\begin{bmatrix} \dot{x}_1 \\ \dot{x}_2 \\ \dot{x}_3 \\ \dot{x}_4 \\ \dot{x}_5 \\ \dot{x}_6 \\ \dot{x}_7 \end{bmatrix}
=
\begin{bmatrix} \dot{\delta} \\ \ddot{\delta} \\ \omega_0 \dot{\psi}_{\mathrm{d}} \\ \omega_0 \dot{\psi}_{\mathrm{f}} \\ \omega_0 \dot{\psi}_{\mathrm{kd}} \\ \omega_0 \dot{\psi}_{\mathrm{q}} \\ \omega_0 \dot{\psi}_{\mathrm{kq}} \end{bmatrix}
= \omega_0
\begin{bmatrix}
 & \frac{1}{\omega_0} & & & & & \\
 & -\frac{K}{J\omega_0} & & & & & \\
 & & -Y_{1\mathrm{d}}R_{\mathrm{a}} & -Y_{4\mathrm{d}}R_{\mathrm{a}} & -Y_{5\mathrm{d}}R_{\mathrm{a}} & -1 & \\
 & & -Y_{4\mathrm{d}}R_{\mathrm{f}} & -Y_{2\mathrm{d}}R_{\mathrm{f}} & -Y_{6\mathrm{d}}R_{\mathrm{f}} & & \\
 & & -Y_{5\mathrm{d}}R_{\mathrm{kd}} & -Y_{6\mathrm{d}}R_{\mathrm{kd}} & -Y_{3\mathrm{d}}R_{\mathrm{kd}} & & \\
 & & 1 & & & -Y_{1\mathrm{q}}R_{\mathrm{a}} & -Y_{3\mathrm{q}}R_{\mathrm{a}} \\
 & & & & & -Y_{3\mathrm{q}}R_{\mathrm{kq}} & -Y_{2\mathrm{q}}R_{\mathrm{kq}}
\end{bmatrix}
\begin{bmatrix} x_1 \\ x_2 \\ x_3 \\ x_4 \\ x_5 \\ x_6 \\ x_7 \end{bmatrix}
+
\begin{bmatrix} \\ \frac{Me}{J} \\ x_6 x_2 + \omega_0 u_{\mathrm{d}} \\ \\ \\ -x_3 x_2 + \omega_0 u_{\mathrm{q}} \\ \\ \end{bmatrix}
+
\begin{bmatrix} & \\ -\frac{1}{J} & \\ & \\ & \omega_0 \\ & \\ & \\ & \end{bmatrix}
\begin{bmatrix} z_1 \\ z_2 \end{bmatrix}
\tag{5.19}
$$

5.3.2 Simplifications

Equations (5.19) can be simplified by neglecting some of the terms, when this is permissible without too much error. A common simplification is to neglect the terms $\dot{\psi}_d$ and $\dot{\psi}_q$ in Eqns. (5.13). The number of equations can now be reduced by two, by eliminating ψ_d and ψ_q. Furthermore, as explained in Section 9.5, the time step of numerical integration can be greatly increased, so that the computer time is drastically reduced.

A further simplification is obtained by omitting the damper winding, thereby reducing the number of equations to three. However the damper winding can be allowed for approximately, by increasing the value of K. If in addition the machine is connected to an infinite bus and the terms containing R_a and $\dot{\delta}$ in Eqns. (5.13) are neglected, Eqns. (5.4) and (5.13) are combined so that

$$\left.\begin{aligned} \omega_0 \psi_d &= -U_m \cos \delta \\ \omega_0 \psi_q &= U_m \sin \delta \end{aligned}\right\} \qquad (5.22)$$

and equations (5.23) and (5.24) with only three state variables are obtained.

A further simplification, which can be justified by Doherty's 'constant-flux-linkage theorem', neglects the rate of change of ψ_f which is proportional to the voltage behind transient reactance (see Section 8.5.2).

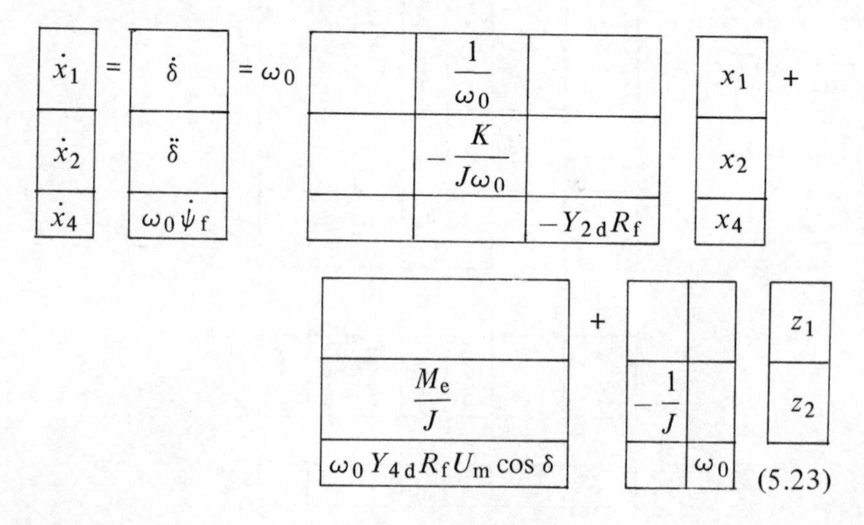

$$\begin{bmatrix} \dot{x}_1 \\ \dot{x}_2 \\ \dot{x}_4 \end{bmatrix} = \begin{bmatrix} \dot{\delta} \\ \ddot{\delta} \\ \omega_0 \dot{\psi}_f \end{bmatrix} = \omega_0 \begin{bmatrix} & \frac{1}{\omega_0} & \\ & -\frac{K}{J\omega_0} & \\ & & -Y_{2d}R_f \end{bmatrix} \begin{bmatrix} x_1 \\ x_2 \\ x_4 \end{bmatrix} + \begin{bmatrix} \\ \frac{M_e}{J} \\ \omega_0 Y_{4d} R_f U_m \cos \delta \end{bmatrix} + \begin{bmatrix} & \\ -\frac{1}{J} & \\ & \omega_0 \end{bmatrix} \begin{bmatrix} z_1 \\ z_2 \end{bmatrix} \qquad (5.23)$$

where

$$M_e = -\frac{U_m Y_{4d} \omega_0 \psi_f}{2} \sin \delta - \frac{U_m{}^2}{4}\{Y_{1q} - Y_{1d}\} \sin 2\delta \qquad (5.24)$$

Hence

$$\dot{x}_1 = x_2$$

$$\left.\begin{aligned} \dot{x}_2 = \frac{-Kx_2}{J} - \frac{U_m{}^2}{4J}(Y_{1q} - Y_{1d}') \sin 2\delta \\ - \frac{U_m Y_{4d}}{2R_f J Y_{2d}} \cdot z_2 \sin \delta - \frac{z_1}{J} \end{aligned}\right\} \qquad (5.25)$$

where

$$Y_{1d}' = Y_{1d} - Y_{4d}{}^2/Y_{2d}$$

Approximations of this kind, and even more drastic ones, were used in the past to determine the transient power-angle curve to which the equal area criterion was applied and also for network analyser studies (see Section 8.5.4); these methods were considered to calculate a swing curve with reasonable accuracy during the first swing but to be too inaccurate for subsequent swings. However that may be, there is now no need, with the availability of the digital computer, to use an inaccurate mathematical model. It is for most purposes admissible to neglect the $\dot{\psi}_d$ and $\dot{\psi}_q$ terms, but the error in leaving out the damper winding is usually not acceptable. Methods of calculating a swing curve from the equations are described in Section 5.6.

The flux linkages were chosen as the state variables in the above derivation because the equations were obtained in a very direct manner and the coefficients retained a physical meaning. Nevertheless, other variables could also have been selected. For problems concerned with a single synchronous machine connected to an infinite bus, the flux linkages are probably the best choice. On the other hand, it may be preferable for a multi-machine system to choose the armature currents i_d and i_q, which appear in the transformations between the reference frames of the generator axis and of the external network. The change from ψ_d, ψ_q, to i_d, i_q, requires only a simple matrix manipulation.

5.4 Calculation of system response and stability

The earlier development of control system theory was concerned mainly with linear systems and a complete technique for calculating their performance was developed. The methods depended on the setting up, using Laplace transform theory, of a transfer function relating an output quantity to an input quantity. The response to a given input signal could then be calculated. Usually a sinusoidal or step-function input function was considered. The linear theory is applicable to those synchronous machine problems, as explained with reference to Table 5.1, for which either speed can be assumed to be constant or the variation relative to a steady condition is small. Even for a relatively simple linear system, the full calculation of the response is a lengthy process and much attention has been given to the development of criteria by which the stability can be assessed without obtaining a complete solution. The criteria are based on the roots of the *characteristic equation*, obtained by equating the denominator of the transfer function to zero.

For a general transient condition in a synchronous machine, the equations are non-linear and the only method of calculating the response is by a step-by-step computation. For complicated systems, particularly when there are multiple controls, the state-space theory provides an effective method of organizing the equations for solution by a computer.

5.4.1 Linear systems

There are three main methods of calculating the response of a linear system when the initial values of $\boldsymbol{x}$ are known.

1. Transfer function method.
2. Eigenvector method.
3. Transition matrix method.

Taking Laplace transforms of Eqn. (5.8),

$$p\bar{\boldsymbol{x}} - \boldsymbol{x}(0) = \boldsymbol{A}\bar{\boldsymbol{x}} + \boldsymbol{B}\bar{\boldsymbol{z}}$$

or

$$\bar{\boldsymbol{x}} = [p\boldsymbol{I} - \boldsymbol{A}]^{-1}\,[\boldsymbol{x}(0) + \boldsymbol{B}\bar{\boldsymbol{z}}]. \tag{5.26}$$

In general $\boldsymbol{z}$ is any function of t, and $\bar{\boldsymbol{z}}$ may introduce additional factors containing p in the denominator. However if the system is *autonomous* ($z = 0$) or if z is an initial step change, the denominator of the transfer function for any x_i is the determinant of $pI - A$ and the characteristic equation is found by equating the determinant to zero.

$$|pI - A| = 0 \qquad (5.27)$$

The roots of Eqn. (5.27) are called *eigenvalues.* The system is stable if the eigenvalues have negative real parts; a conclusion that does not depend on the magnitude or the point of application of the step input or on the initial conditions. Since any root p_i obeys Eqn. (5.26), the following expanded set of equations holds.

$$\begin{bmatrix} A_{11} - p_1 & A_{12} & \ldots \; A_{1n} \\ A_{21} & A_{22} - p_i & A_{2n} \\ \ldots & & \ldots \\ A_{n1} & A_{n2} & A_{nn} - p_1 \end{bmatrix} \begin{bmatrix} x_1 \\ x_2 \\ \ldots \\ x_n \end{bmatrix} = \begin{bmatrix} x_1(0) \\ x_2(0) \\ \ldots \\ x_n(0) \end{bmatrix} \qquad (5.28)$$

Transfer function method

For the transfer function method, any x_i due to a given $x_j(0)$ can be found by eliminating all values of x other than x_i and putting all $x(0)$ other than $x_j(0)$ equal to zero.

$$\bar{x}_i = f(p)x_j(0). \qquad (5.29)$$

The solution is obtained by taking the inverse transform of $f(p)$. For a simple system the transfer function can be obtained more easily from the physical relationships without deriving the state-space equations. For complicated systems, particularly those with multiple controls, the other two methods are preferable, because they lead to a better organised programme for computation.

Eigenvector method

The solution for x_i is of the form

$$x_i = E_{1i}e^{-jp_1t} + E_{2i}e^{-jp_2t} + \ldots + E_{ni}e^{-jp_nt} \qquad (5.30)$$

where each term is a solution of Eqn. (5.28) with zeros on the right hand side. For a given p_j the solution of the homogeneous equations can only yield the ratios of the n variables.

A set of values $(E_{j1}, E_{j2}, \ldots E_{jn})$ is called an *eigenvector* corresponding to the eigenvalue p_j. To complete the solution the initial values $x(0)$ must be introduced to determine the absolute values of the coefficients of Eqn. (5.30).

Transition matrix method

The solution can be computed by the following formula, in which Φ is called the *transition matrix* corresponding to A.

$$\boldsymbol{x}(t) = [\Phi(t)]\boldsymbol{x}(t_0) + \int_{t_0}^{t_f} [\Phi(t-\tau)]\boldsymbol{B} \cdot \boldsymbol{z}(\tau)\, d\tau. \tag{5.31}$$

The above brief discussion of the methods of solution is only intended to indicate what alternative methods are available. A fuller treatment is given in many books on Control System Theory.

5.4.2 Non-linear systems

The solution of the equations of a non-linear system as complicated as a synchronous machine, can only be obtained by a numerical integration method, for which the state-space form is the most convenient. Some notes on alternative methods are given in Section 5.6.

Stability must be defined more carefully than for a linear system, since it depends on both the input and the initial conditions, as well as on the system equations. The stability of a synchronous machine system is usually judged from the manner of variation with time of the transient rotor angle δ. This *swing curve* can be computed for a specified large disturbance, applied when the machine is operating at a given steady condition. Fig. 5.2, curve A, shows how the rotor angle δ varies with time when a machine, while running at a steady angle δ_0, is subjected to a short-circuit which is cleared after time t_{1a}. Curve A is an example of a stable condition, for which δ returns to its original value after a damped oscillation. Curve B, obtained with an increased switching time t_{1b}, shows an unstable condition, for which δ increases while the machine pulls out of step.

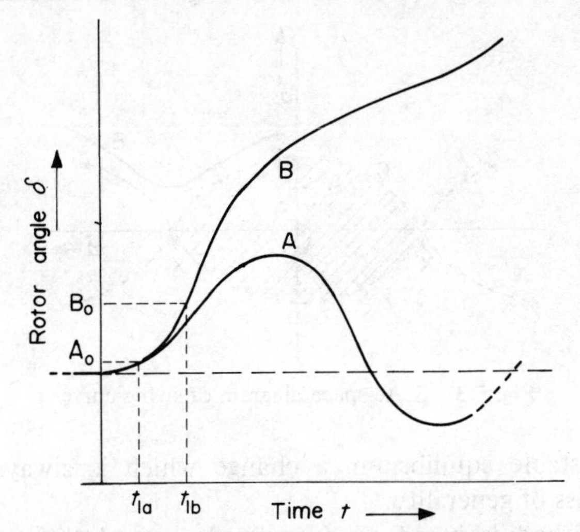

Fig. 5.2 Swing curve of synchronous machine subjected to temporary short-circuit. curve A – stable; curve B – unstable.

Fig. 5.3 is a two-dimensional state-space (or phase-plane) diagram showing how $\dot{\delta}$ varies with δ for the same condition as in Fig. 5.2. To avoid confusion the curves start at the points A_0 and B_0 at which the short-circuit is cleared. Curve A is the stable curve, which spirals to a steady synchronous condition, and the system is said to be *asymptotically stable.* Curve B, which starts at the point B_0 for which δ_0 is greater, settles to a periodic condition during which the slip oscillates with the machine out of synchronism. With a still longer switching time the slip may increase indefinitely (curve not shown) and the machine runs away. Curve C is the boundary, known as the *separatrix*, between the stable and unstable regions. Points of stable equilibrium like X occur at intervals of 2π along the δ-axis. The points of unstable equilibrium like X′, where the separatrix meets the δ-axis, also occur at intervals of 2π.

Liapunov's direct method

The so-called direct method of Liapunov provides a method of assessing the stability of a system without solving the differential equations. It applies to an autonomous system, like the synchronous machine illustrated by Fig. 5.3, for which the control variables are zero. In applying the method, the origin is transferred to a

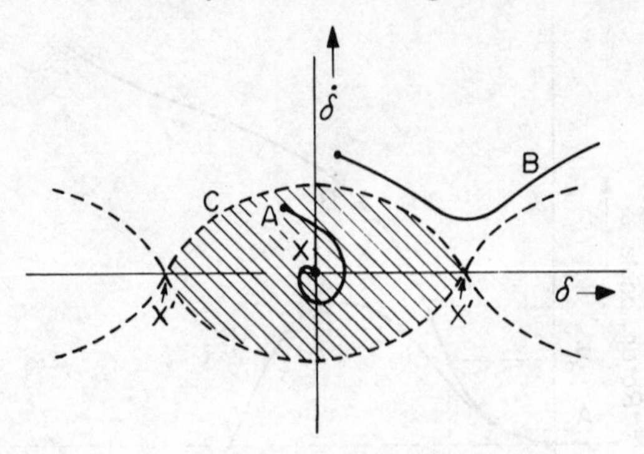

Fig. 5.3 State-space diagram of swing curves.

point of stable equilibrium, a change which is always possible without loss of generality.

The method is based on Liapunov's second theorem, which states that if a *positive definite* function V (a function of the state variables) can be found over a region such that its time-derivative is *negative semi-definite*, the system is stable over that region. A positive (or negative) definite function is one which is everywhere positive (or negative) over the region and zero at the origin. A positive (or negative) semi-definite function is one which is greater than (less than) or equal to zero everywhere over the region and zero at the origin.

The Liapunov method applies to both linear and non-linear systems, but since there are many simpler criteria for the stability of linear systems, its main usefulness is likely to be to non-linear systems. However, the concept and its interpretation on the phase-plane diagram can be simply illustrated by considering the linear mechanical system for which the state equations are Eqns. (5.10) with $z = 0$. The energy function

$$V = b^2 x_1{}^2 + x_2{}^2$$

obtained as the sum of the potential energy in the spring and the kinetic energy in the mass, is a possible Liapunov function. The derivative is

$$\dot{V} = 2b^2 x_1 \dot{x}_1 + 2x_2 \dot{x}_2 = -2ax_2{}^2$$

which is negative semi-definite if $a > 0$. This stability criterion agrees with that obtained by the conventional method.

The method is illustrated by the phase-plane diagram of Fig.

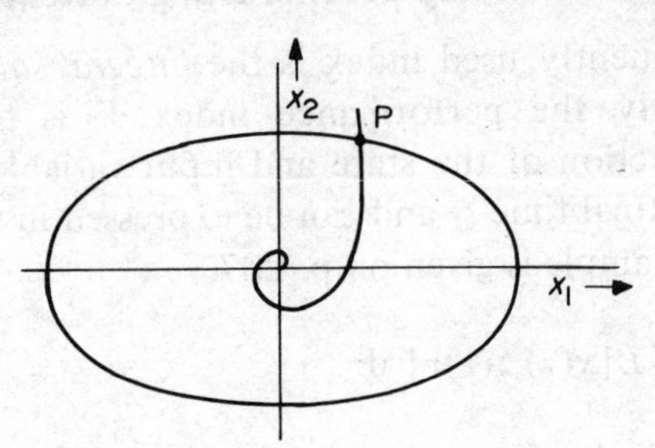

Fig. 5.4 Two dimensional state-space diagram.

5.4. The curve showing the relationship between x_1 and x_2 for a given constant value of V is an ellipse and P is the point on the trajectory which lies on the ellipse. Since $\dot{V}$ is negative semi-definite at this point, the trajectory spirals to the origin as shown. Thus for this simple example the Liapunov criterion is equivalent to a statement that the energy must decrease continuously to zero.

The linear system is stable for initial values anywhere in the plane, if the criterion is satisfied, but a non-linear system would normally be stable only in a limited region. For a non-linear system of higher order it becomes increasingly difficult to find a Liapunov function and, when found, the calculated region of stability is usually smaller than the practical region, so that the result is pessimistic. Many papers have been written applying the method to the synchronous machine, but most authors have used over-simplified equations and so far it has not been widely used in power system computations [48].

5.5 Optimization. Performance indices

When a method of solving a set of equations has been established, its use can be extended to optimize some property of the system which can be expressed as a function of the variables. Such a function is called a *performance index* or *cost function* [46]. The optimization can be carried out in relation to any of the factors which determine the operation of the system. Generally it is carried out in relation to the input variables $\boldsymbol{z}$ in Eqn. (5.6).

For a synchronous machine, for which δ (see Fig. 5.2) is an important quantity for assessing its stability, a possible index is the time for δ to fall to within a given amount of its final value.

Another frequently used index is the *integral squared error* of δ. More generally the performance index V is the integral of a quadratic function of the state and input variables from an initial time t_0 to a final time t_f and can be expressed in the form of Eqn. (5.32). An example is given on p. 207.

$$V(\boldsymbol{x},t_0) = \int_{t_0}^{t_f} L[\boldsymbol{x}(\tau),\boldsymbol{z}(\tau)\tau]\ \mathrm{d}\tau. \tag{5.32}$$

The *principal of optimality* states that any portion of an optimal trajectory, starting at time t and ending at time t_f, is itself optimal over that interval. Hence the function $V(\boldsymbol{x},t)$, in which t_0 is replaced by t, is also optimal. Let the optimal value be

$$S(\boldsymbol{x},t) = \min_{\boldsymbol{z}(t)}.\left\{\int_{t}^{t_f} L[\boldsymbol{x}(\tau),\boldsymbol{z}(\tau),\tau]\ \mathrm{d}\tau\right\}$$

It is assumed that S exists and is differentiable. The change in S between instants t and $(t + \Delta t)$ is

$$\Delta S = S(\boldsymbol{x} + \Delta\boldsymbol{x},\ t + \Delta t) - S(\boldsymbol{x},t) = \min_{\boldsymbol{z}(t)}.\left[\int_{t}^{t+\Delta t} L[\boldsymbol{x}(\tau),\boldsymbol{z}(\tau),\tau]\ \mathrm{d}\tau\right] \tag{5.33}$$

When Δt tends to zero, Eqn. (5.33) may be approximated to

$$\Delta S = -\min_{\boldsymbol{z}(t)}.\{L[\boldsymbol{x}(t),\boldsymbol{z}(t),t]\ \Delta t\}$$

which also equals

$$\left[\frac{\partial S}{\partial \boldsymbol{x}}\right]^{\mathrm{T}} \cdot \Delta\boldsymbol{x} + \frac{\partial S}{\partial t} \cdot \Delta t \tag{5.34}$$

but $[\partial S/\partial \boldsymbol{x}]^{\mathrm{T}} \cdot \Delta\boldsymbol{x}$ can be put inside the minimizing bracket because this quantity is already minimized. When $\dot{\boldsymbol{x}}$ is substituted from Eqn. (5.6), equation (5.34) becomes

$$-\frac{\partial S}{\partial t} = \min_{\boldsymbol{z}(t)}.\left\{L[\boldsymbol{x}(t),\boldsymbol{z}(t),t] + \left[\frac{\partial S}{\partial \boldsymbol{x}}\right]^{\mathrm{T}} \dot{\boldsymbol{x}}\right\} \tag{5.35}$$

$$= \min_{\boldsymbol{z}(t)}.\left\{L[\boldsymbol{x}(t),\boldsymbol{z}(t),t] + \left[\frac{\partial S}{\partial \boldsymbol{x}}\right]^{\mathrm{T}} \boldsymbol{F}[\boldsymbol{x}(t),\boldsymbol{z}(t)]\right\}$$

The superscript T indicates the transpose of a matrix.

Equation (5.35) is known as Bellman's equation. Several methods have been proposed for translating it into a practical solution, for example by using the technique of dynamic programming which determines the input vector **z** as a function of time. However, it would be more useful to obtain **z** as a function of **x**, so that feedback regulators could be designed. The following few paragraphs indicate how this may be achieved for a linear system.

5.5.1 Linear systems

A large saving in complexity is realized when in the case of linear systems, described by Eqn. (5.8) the performance index may be chosen such that

$$S(\boldsymbol{x},t) = \min_{\boldsymbol{z}(t)}. [\tfrac{1}{2}\int_{t0}^{t\mathrm{f}} (\boldsymbol{x}^{\mathrm{T}}\boldsymbol{Q}\boldsymbol{x} + \boldsymbol{z}^{\mathrm{T}}\boldsymbol{R}\boldsymbol{z})\, \mathrm{d}t] \tag{5.36}$$

where

$\boldsymbol{Q} = n \times n$ weighting matrix

$\boldsymbol{R} = m \times m$ weighting matrix

Assume for a trial solution that

$$S(\boldsymbol{x},t) = \tfrac{1}{2}\boldsymbol{x}^{\mathrm{T}}\boldsymbol{K}(t)\boldsymbol{x} \tag{5.37}$$

where $\boldsymbol{K}(t)$ is a symmetric matrix.

Then

$$\frac{\partial S}{\partial \boldsymbol{x}} = \boldsymbol{K}(t)\boldsymbol{x} \tag{5.38}$$

and

$$\frac{\partial S}{\partial t} = \tfrac{1}{2}\boldsymbol{x}^{\mathrm{T}}\dot{\boldsymbol{K}}(t)\boldsymbol{x} \tag{5.39}$$

Substituting from Eqns. (5.8), (5.36), (5.38) and (5.39) into Eqn. (5.35) yields

$$-\tfrac{1}{2}\boldsymbol{x}^{\mathrm{T}}\dot{\boldsymbol{K}}(t)\boldsymbol{x} = \min_{\boldsymbol{z}(t)}. \left\{\tfrac{1}{2}\boldsymbol{x}^{\mathrm{T}}\boldsymbol{Q}\boldsymbol{x} + \tfrac{1}{2}\boldsymbol{z}^{\mathrm{T}}\boldsymbol{R}\boldsymbol{z} + [\boldsymbol{K}(t)\boldsymbol{x}]^{\mathrm{T}}[\boldsymbol{A}\boldsymbol{x} + \boldsymbol{B}\boldsymbol{z}]\right. \tag{5.40}$$

The right-hand side of Eqn. (5.40) is evaluated by differentiating with respect to **z** and equating the result to zero, to yield

$$\boldsymbol{R}\boldsymbol{z}(t)_{\mathrm{min.}} = -[\boldsymbol{K}(t) \cdot \boldsymbol{x}(t)]^{\mathrm{T}}\boldsymbol{B}$$

or

$$z(t)_{\text{min.}} = -\boldsymbol{R}^{-1}\boldsymbol{B}^{\mathrm{T}}\boldsymbol{K}(t)\boldsymbol{x}(t) \tag{5.41}$$

Eqn. (5.41) states the new feedback law between $\boldsymbol{x}(t)$ and $z(t)_{\text{min}}$ but the required value of $\boldsymbol{K}(t)$ is still unknown. This is found by substituting the minimum value of z into Eqn. (5.40) and simplifying to find that

$$\dot{\boldsymbol{K}}(t) = \boldsymbol{K}(t)\boldsymbol{B}\boldsymbol{R}^{-1}\boldsymbol{B}^{\mathrm{T}}\boldsymbol{K}(t) - \boldsymbol{Q} - \boldsymbol{A}^{\mathrm{T}}\boldsymbol{K}(t) - \boldsymbol{K}(t)\boldsymbol{A}, \tag{5.42}$$

which is called the Riccati equation.

The value of $\boldsymbol{K}(t)$ is found by numerical backward integration of Eqn. (5.42) starting from $t = t_{\mathrm{f}}$ when $\boldsymbol{K}(t_{\mathrm{f}}) = 0$. This solution of $\boldsymbol{K}$ as a function of time, may however be precalculated and stored for call-up when necessary since the calculation does not involve $\boldsymbol{x}(t)$. Examples of optimal control studies on synchronous machines are described in Section 9.4.

The explanations in Sections 5.3 to 5.5 of the methods using state variable equations are very brief. They are included here for the benefit of machine and power system engineers who are not familiar with these methods. The authors have found that the books and papers on the subject usually assume that the reader is already familiar with modern control theories.

5.6 Computational techniques for transient studies

In any power system, it is important to know how the system will perform during certain transient conditions and over the years considerable effort has gone into the development of methods to predict the performance. The main variable of interest in a synchronous machine is the rotor angle δ, since its variation as a function of time, called the swing curve, gives a direct indication of its stability. An analogue computer may be used to determine the performance, particularly when the purpose is to investigate a new problem, but the commonest method uses a digital computer, which calculates the variables by a numerical step-by-step integration of the non-linear differential equations.

An earlier form of the step-by-step method used a network analyser to represent the conditions at any instant. From readings taken on the analyser, the setting of the analyser for the next step was determined by a manual calculation. Later on, analogue devices were developed to carry out the intermediate calculation. The method was based on what would now be regarded as excessive simplification.

The computer time required for a digital computation is an important factor in determining the method and the degree of simplification of the equations used. For calculations for a multi-machine system, in which each generator has a voltage regulator and a governor, the computer time can be several times the real time of the transient variation. Thus the choice of the computing technique is important.

The computations described above are *off-line* calculations. An *on-line* calculation is one where the computer is used to monitor and control the process [59].

In Fig. 5.5 the block labelled *model* is that portion of the computer associated with calculating, from a mathematical model of the *plant*, the response to a certain *input*. Likewise, the block labelled *adaptive control law* is that portion of the computer associated with calculating the feedback signal from the *error* between the calculated model output and the actual plant output. With suitable equations to represent the control law, the plant is forced to follow almost exactly the behaviour predicted by the model. A mathematical model may therefore be chosen to force any desired response onto the plant. On-line methods are coming into use for controlling the steady load distribution of a power system, but their use for transient conditions is much more difficult because of the rapid variation.

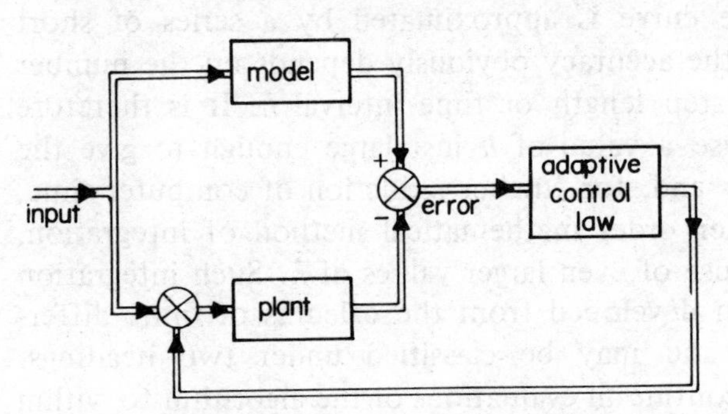

Fig. 5.5 Model reference adaptive control for a turbo-generator.

5.6.1 Step-by-step integration

When $\boldsymbol{z}$ is known, either as an independent function of time or as a function of $\boldsymbol{x}$, the differential equation for $\boldsymbol{x}$ has the general non-linear form of Eqn. (5.43) and cannot be integrated analytically.

$$\dot{\boldsymbol{x}} = \boldsymbol{f}(\boldsymbol{x},t) \tag{5.43}$$

Let x_n be the value of a typical variable at time t_n. The principle of the step-by-step method is that, when the value of x_n at time t_n is known, the equations can be used to calculate the derivative $\dot{x}_n$, and hence to determine the value at the succeeding instant $t_{n+1} = t_n + h$. The solution values at t_{n+1} are approximated from values at the beginning of the step, possibly from values in earlier steps, and from slope evaluations which depend in form and combination on the particular method used. Errors are propagated from step to step and the error build-up depends largely on the step length. From the outset then, there is a conflict between the adequate control of error growth, on the one hand, and the overall computing time required to achieve it, on the other.

A simple method is to say that

$$x_{n+1} = x_n + h\dot{x}_n \tag{5.44}$$

The method is self-starting, since the derivatives at time $t = 0$ depend only on the values of $\boldsymbol{x}_0$ at that instant and will therefore integrate satisfactorily following a discontinuity occurring at the beginning on any step. It is of first order accuracy, since it computes the slope $\boldsymbol{f}(\boldsymbol{x},t)$ only once during each step.

Any numerical integration method is of course only approximate, because the curve is approximated by a series of short straight lines and the accuracy obviously depends on the number of steps, i.e. the step length or time interval h. It is therefore advantageous to use a value of h just large enough to give the necessary accuracy and, for further reduction of computer time, to develop a higher order mathematical method of integration, which allows the use of even larger values of h. Such integration methods have been developed from the calculus of finite difference polynomials and may be classified under two headings, namely those that confine all evaluations of the algorithm to within each integration step, on the one hand, and those that use solution information available in previous steps, on the other. Of the two methods most frequently used in power system studies, the

Runge-Kutta algorisms fall within the first group of methods, and the Predictor-Corrector algorithms into the second group.

The most commonly used version of Runge-Kutta algorithms [27] is one in which four evaluations of the slope are made within the duration of the integration step and from these, a mean weighted slope S is computed of fourth order accuracy. [see Eqn. (5.45)]. The algorithm which appears in Eqn. (5.45) stems from a truncated infinite series expansion of x over the interval h. Eqn. (5.44) with $\dot{x}_n$ replaced by S is used to calculate x_{n+1}.

$$x_{n+1} = x_n + \frac{K_0}{6} + \frac{K_1}{3} + \frac{K_2}{3} + \frac{K_3}{6}$$

where

$$\left.\begin{aligned} K_0 &= hf(x_n, t_n) \\ K_1 &= hf\left(x_n + \frac{K_0}{2}, t_n + \frac{h}{2}\right) \\ K_2 &= hf\left(x_n + \frac{K_1}{2}, t_n + \frac{h}{2}\right) \\ K_3 &= hf(x_n + K_2, t_n + h) \end{aligned}\right\} \tag{5.45}$$

Also within the family of Runge-Kutta algorithms is the fifth order Kutta-Merson set of equations (5.46) which allow an even larger value of h to be used,

$$\left.\begin{aligned} K_0 &= \frac{h}{3} \cdot f(x_n, t_n) \\ K_1 &= \frac{h}{3} \cdot f\left(x_n + K_1, t + \frac{h}{3}\right) \\ K_2 &= \frac{h}{3} \cdot f\left[x_n + \tfrac{1}{2}(K_1 + K_2), t + \frac{h}{3}\right] \\ K_3 &= \frac{h}{3} \cdot f\left[x_n + \tfrac{1}{8}(3K_1 + 9K_3), t + \frac{h}{2}\right] \\ K_4 &= \frac{h}{3} \cdot f[x_n + \tfrac{1}{2}(3K_1 - 9K_3 + 12K_4), t + h] \\ x_{n+1} &= x_n + \frac{K_1}{2} + 2K_4 + \frac{K_5}{2}. \end{aligned}\right\} \tag{5.46}$$

Whereas the Runge-Kutta routine starts each step afresh, and utilizes only slopes within each step, the predictor-corrector methods make use of the variable and its first derivative obtained in previous steps to predict at t_n the value of x_{n+1} and then use a corrector formula to improve the accuracy of the estimate. The accuracy depends upon the number of previous values of x and $\dot{x}$ which are available and on the relative significance attached to each one. Geometrically the process may be illustrated by Fig. 5.6.

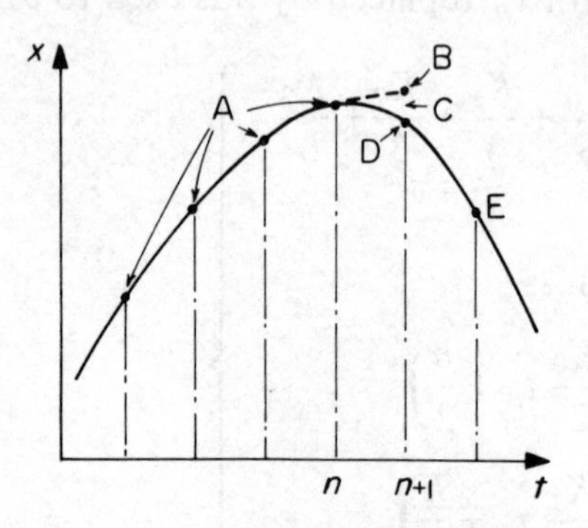

Fig. 5.6 Geometric interpretation of one step in a predictor-corrector integration process.
A – Previous points used in both predictor and corrector.
B – Predicted point.
C – Prediction successively corrected towards true solution.
D – Final point.
E – True solution.

An example of a predictor-corrector pair is given by the following Adams-Bashforth formulas [31] of fifth order accuracy:

$$\left.\begin{aligned}\bar{x}_{n+1} &= x_n + \frac{h}{24}(55\dot{x}_n - 59\dot{x}_{n-1} + 37\dot{x}_{n-2} - 9\dot{x}_{n-3}) \\ x_{n+1} &= x_n + \frac{h}{24}(9\dot{\bar{x}}_{n+1} + 19\dot{x}_n - 5\dot{x}_{n-1} + \dot{x}_{n-2})\end{aligned}\right\} \quad (5.47)$$

where $\bar{x}_{n+1}$ is the predicted value and x_{n+1} the corrected value of x at t_{n+1}.

Because the prediction depends on previous values of the integral, the method is not inherently self-starting. When there is an abrupt change in the function being integrated, the next predicted value which is based on the knowledge of several previous values, may be considerably in error. The corrector is then called upon to eradicate an unduly large error which may not be possible. This loss of accuracy may be overcome by increasing the corrector iterations within a step, or decreasing the step length, or temporarily replacing the predictor-corrector method by a Runge-Kutta method for a few steps as a starter only. Care should be taken to use a starter of the same accuracy order as the predictor-corrector. Further examples of predictor-corrector methods appear in [31 and 65].

5.6.2 Comparison of Runge-Kutta and predictor-corrector methods

The advantages of the Runge-Kutta methods are mainly that they are self-starting and have high accuracy. However, the disadvantage is that, as in the case of Eqn. (5.46), the slope has to be computed five times within each time step, thereby using excessive computer time; accuracy diminishes rapidly as the step length increases and the method has limited stability.

Predictor-corrector methods have the advantage of high accuracy and stability and require calculation of the slope only once per step as in Eqn. (5.47). Their disadvantage however is that they cannot integrate through discontinuities.

Chapter Six
Automatic Control of Synchronous Machines

6.1 General

The terminal voltage and speed of a synchronous generator are two important quantities on which the operation depends, and special means are necessary to control them if the best operating conditions are to be obtained. The voltage and speed of a generator operating by itself can be wholly controlled by an excitation regulator actuated by a voltage feedback and a turbine governor actuated by a speed feedback. When a generator is connected to a large power system, however, its voltage and speed are determined much more by the voltage and frequency of the system than by the generator conditions. Nevertheless voltage regulators and speed governors play a large part in the operation of the system and are always provided on a large generator. The voltage of the system is to some extent flexible, but the main effect of a voltage regulator is to control the reactive current. On the other hand the system frequency is nearly constant and the speed changes are small, so that the speed governor is in effect a load controller. A speed signal is not necessarily the best method of actuation, although the governor is required to perform the speed control function in an emergency in order to prevent the generator from running away.

Feedback control devices are often used with synchronous motors in order to control the power factor and increase the pull-out torque. For a generator operating in isolation, a regulator and a governor are essential. The generator in a power system is, however, of more practical importance than an isolated generator and the main emphasis in the following chapters is on such systems. A well designed control system can not only improve the steady performance, it can also improve the stability.

6.2 Excitation control of a.c. generators

In principle, a voltage signal is taken from a rectifier bridge connected to the generator output. After comparison with a reference voltage, the error signal is processed by the regulator, the transfer function of which is $K(p)$, and applied to the generator field winding. For the purpose of analysis, the exciter is regarded as a part of the regulating system, in which it serves as a power amplifier. Both the practical construction of the regulator and its transfer function can take many forms. The exciter may be a d.c. machine or an a.c. machine with rectifiers or thryistors and the rest of the power amplification may be provided by an amplidyne, magnetic amplifiers or electronic amplifiers. Alternatively the amplifier may be all electronic, without a machine exciter.

The regulator transfer function $K(p)$, which determines the performance of the system, can be classified into three basic types, viz.

(a) proportional type, $$K(p) = \frac{K}{1 + \tau p}$$

(b) integrator type, $$K(p) = \frac{K}{p(1 + \tau p)}$$

(c) derivative type, $$K(p) = \frac{Kp}{1 + \tau p}$$

A practical regulator generally has a transfer function which combines the features of the three basic types and it also has additional stabilizing feedback loops within the regulator.

Fig. 6.1 shows the block diagram of a typical regulator. Details of the regulator, which was used with a 37.5 MVA machine, have been published in [29] and the system has been used in several subsequent theoretical investigations. The regulator equations are given in state variable form in Eqn. (6.2) in which K_{m1}, etc. are constant bias values and z_1 and z_2 have values given in Eqn. (6.1).

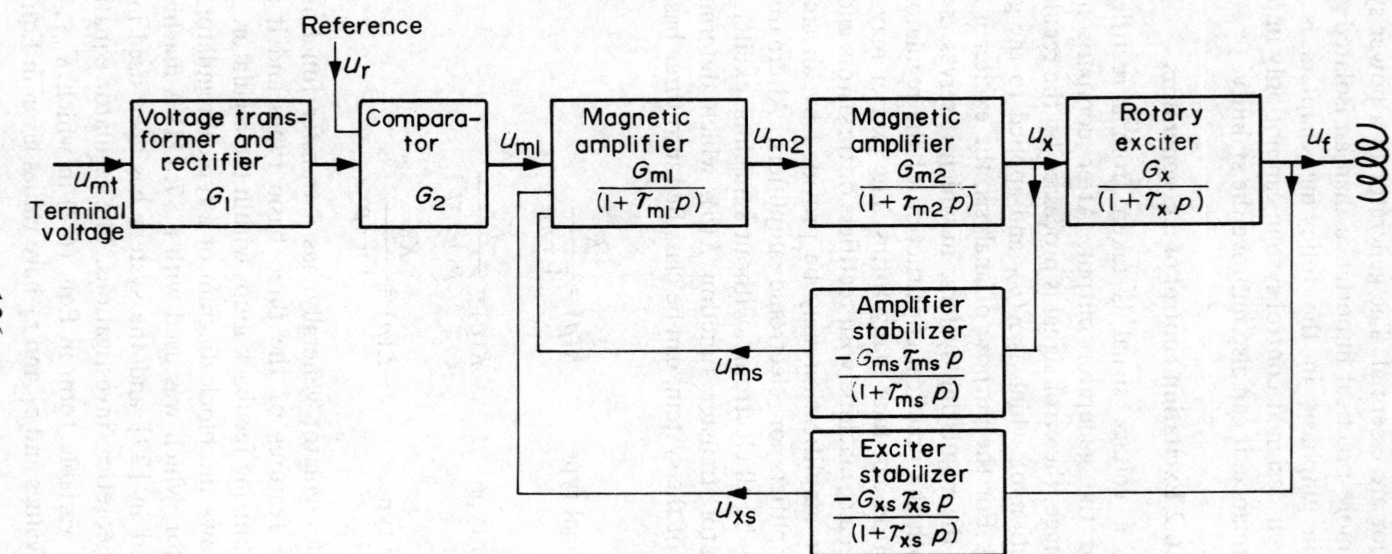

Fig. 6.1 Block diagram of a voltage regulator for a 37.5 MVA generator.

$$
\begin{bmatrix} \dot{x}_1 \\ \dot{x}_2 \\ \dot{x}_3 \\ \dot{x}_4 \\ \dot{x}_5 \end{bmatrix}
=
\begin{bmatrix} \dot{u}_{m2} \\ \dot{u}_{x} \\ \dot{u}_{f} \\ \dot{u}_{ms} \\ \dot{u}_{xs} \end{bmatrix}
=
\begin{bmatrix}
-1/\tau_{m1} & & & G_{m1}/\tau_{m1} & G_{m1}/\tau_{m1} \\
G_{m2}/\tau_{m2} & -1/\tau_{m2} & & & \\
 & G_{x}/\tau_{x} & -1/\tau_{x} & & \\
\dfrac{-G_{ms}G_{m2}}{\tau_{m2}} & \dfrac{G_{ms}}{\tau_{m2}} & & -1/\tau_{m} & \\
 & \dfrac{-G_{xs}G_{x}}{\tau_{x}} & G_{xs}/\tau_{x} & & -1/\tau_{xs}
\end{bmatrix}
\begin{bmatrix} x_1 \\ x_2 \\ x_3 \\ x_4 \\ x_5 \end{bmatrix}
+
$$

$$
\begin{bmatrix} K_{m1}/\tau_{m1} \\ K_{m2}/\tau_{m2} \\ K_{x}/\tau_{x} \\ \dfrac{-G_{ms}K_{m2}}{\tau_{m2}} \\ \dfrac{-G_{xs}K_{x}}{\tau_{x}} \end{bmatrix}
+
\begin{bmatrix}
\dfrac{G_{m1}G_{2}}{\tau_{m1}} & \dfrac{G_{m1}G_{1}G_{2}}{\tau_{m1}} \\
 & \\
 & \\
 & \\
 &
\end{bmatrix}
\begin{bmatrix} z_1 \\ z_2 \\ \\ \\ \end{bmatrix}
\qquad (6.2)
$$

For the purpose of numerical computations, the terminal voltage u_{mt} is obtained as follows from the axis components of voltage

$$u_{mt} = \sqrt{(u_d^2 + u_q^2)}$$

$$\begin{bmatrix} z_1 \\ z_2 \end{bmatrix} = \begin{bmatrix} u_r \\ u_{mt} \end{bmatrix} \tag{6.1}$$

The elements of the regulator are assumed to be linear in deriving Eqn. (6.2). In practice, the voltage of any amplifier is limited between upper and lower boundaries.

6.3 Quadrature field winding. The divided-winding-rotor generator

6.3.1 Quadrature field winding

Instead of only the one field winding (Fig. 4.2), a synchronous machine may have two field windings represented by coils FD and FQ in Fig. 6.2. With an angle regulator feeding the quadrature-axis

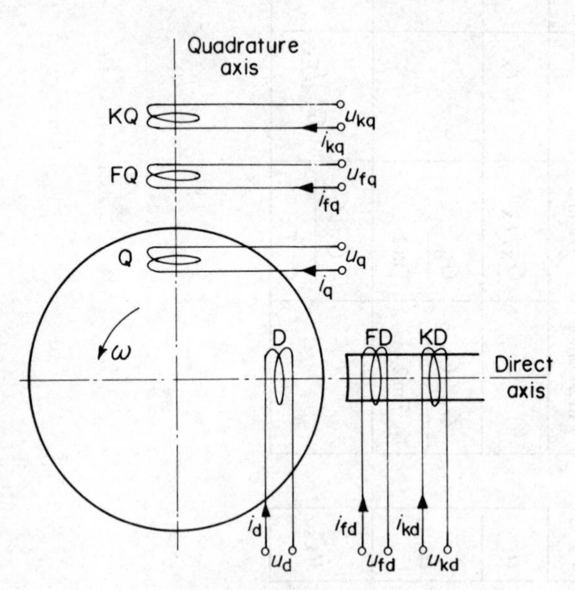

Fig. 6.2 Diagram of a primitive synchronous machine with two field coils and two damper coils.

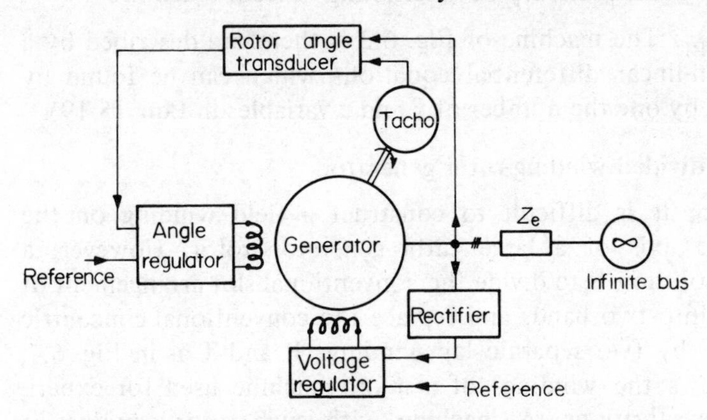

Fig. 6.3 Diagram of a synchronous machine with a regulator feeding a field winding on each axis.

field and a voltage regulator feeding the direct-axis field, as illustrated in Fig. 6.3, the resultant field m.m.f. is positioned with respect to the rotor in such a way that the stability of the machine no longer depends on rotor acceleration or deceleration, [44].

The equations for the coils in Fig. 6.2 are similar to those in Section 4.6 except that the subscript f is replaced by fd. Equations (6.3) and (6.4) are needed to allow for the presence of the FQ coil.

$$u_{fq} = R_{fq} i_{fq} + \dot{\psi}_{fq} \tag{6.3}$$

$$\begin{bmatrix} \psi_{fq} \\ \psi_q \\ \psi_{kq} \end{bmatrix} = \begin{bmatrix} L_{mq} + L_{fq} & L_{mq} & L_{mq} \\ L_{mq} & L_{mq} + L_a & L_{mq} \\ L_{mq} & L_{mq} & L_{mq} + L_{kq} \end{bmatrix} \begin{bmatrix} i_{fq} \\ i_q \\ i_{kq} \end{bmatrix} \tag{6.4}$$

Because of the symmetry of the new field winding, the equations for all the q-axis coils are similar to those for the corresponding d-axis coils and the q- and d-axis equivalent circuits are similar. Some new parameters such as $T_q{}'$, $T_{q0}{}'$ and $X_q{}'$ are defined for the q-axis, in the same way as $T_d{}'$, $T_{d0}{}'$ and $X_d{}'$ were defined in Section 4.6.

In comparison with Eqns. (5.11) and (5.12) there now exists the additional state variable $\omega_0 \psi_{fq}$ and the additional input variable u_{fq}. The machine of Fig. 6.2 is therefore described by a set of non-linear differential equations which can be found by increasing by one the number of x and z variables in Eqn. (5.19).

6.3.2 The divided-winding-rotor generator

In practice it is difficult to construct a field winding on the quadrature axis of a large turbo-generator rotor. However, a practical solution is to divide the conventional slot arrangement of the rotor into two bands and replace the conventional concentric winding F by two separate lap windings R and T as in Fig. 6.4, which shows the windings of a micro-machine used for experimental investigations. A machine with such rotor windings is referred to as a *divided-winding-rotor (d.w.r.)* machine and it has a performance quite different from the *conventionally wound rotor (c.w.r.)* machine. The arrangement makes a sound mechanical construction possible and also gives a favourable combination of the two component field currents. Further practical details of a d.w.r. machine and its regulators appear in Refs. [43 and 47].

Neither the R nor the T coil axis coincides with either the direct or quadrature axis and therefore, due to saliency, the axis of the flux wave produced by each coil does not coincide with the respective coil axis. The flux produced by a particular coil may be found by resolving the coil m.m.f. into two components along the direct and quadrature axes, then finding and adding vectorially the flux produced by each m.m.f. component. The relationship between the currents in the R and T windings and those in the FD and FQ windings of Fig. 6.2 is given by the following transformation, based on the equivalence of the resultant m.m.f.s.

$$\begin{bmatrix} i_{fd} \\ i_{fq} \end{bmatrix} = \begin{bmatrix} \cos\phi_t & N\cos\phi_r \\ -\sin\phi_t & N\sin\phi_r \end{bmatrix} \begin{bmatrix} i_t \\ i_r \end{bmatrix} \tag{6.5}$$

where ϕ_r and ϕ_t are the angles defining the positions of coils R and T and N is their turns ratio. The d.w.r. machine can be analysed by using the equations for the machine of Fig. 6.2, in conjunction with the transformation of Eqn. (6.5).

The application of a synchronous machine with two field windings is discussed further in Section 7.6.

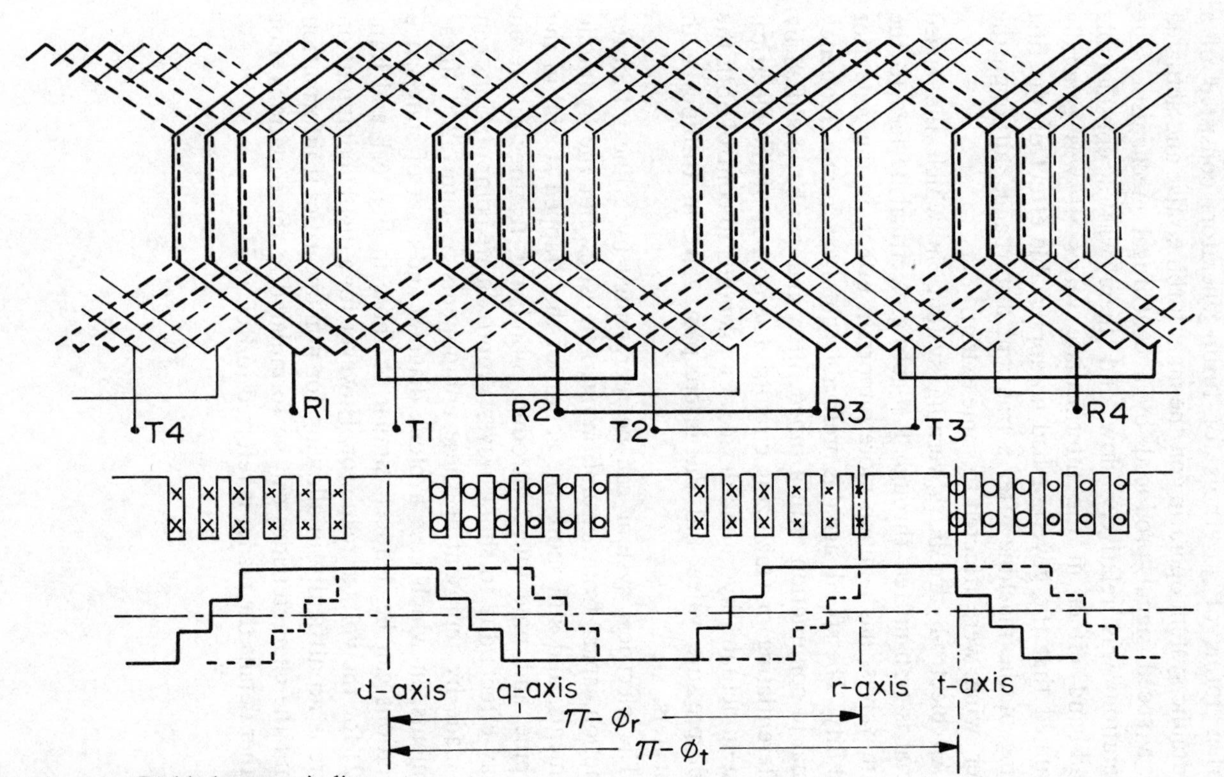

Fig. 6.4 Divided rotor winding.

6.4 Speed governors

Earlier forms of governors for turbo-generators consisted of a hydraulic system operating on the main turbine valve and actuated by a speed signal produced by a centrifugal mechanism. The operation was relatively slow and there was a considerable dead-band effect in the intermediate valves. It was generally assumed that the governor had no appreciable effect on the first swing of the machine after a transient disturbance, although the later swings were affected to some extent.

An electro-hydraulic governing system, in which the speed signal is transmitted through fast-acting electrical devices to the main valves, has a much more important effect on the generator stability. The speed signal is preferably obtained from a frequency signal in conjunction with a tuned filter, and can be modified by differentiating or stabilizing circuits. The output of the electrical system actuates hydraulic devices or solenoids to control both the high pressure valve and the interceptor valve in turbines with reheaters.

The electrical system can be represented by a linear transfer function, apart from saturation characteristics, but the hydraulic elements, and still more the action of the 'trapped steam' in the turbine and pipe lines, introduce non-linear features and even delay effects due to shock waves. Approximate computations can be made by representing these components by equivalent linear elements, in which case a block diagram and a set of state-space equations can be deduced similar to those in Fig. 6.1 and Eqn. (6.2) for the voltage regulator. However for a digital computation there is no difficulty in allowing for non-linearities if information about the characteristics can be formulated in the equations. For further details the reader is referred to Ref. [68].

Chapter Seven
A.C. Operation of Synchronous Machines

7.1 Steady operation of the synchronous machine at synchronous speed

In Section 3.4 the phasor diagram and phasor equation of a three-phase synchronous machine were derived by considering the rotating m.m.f. wave set up by the armature currents. In the present section the same result is deduced as a special case from the general equations.

During normal steady a.c. operation, the speed of the machine is the constant synchronous speed ω_0. The field voltage and current are constant, the damper currents are zero, and the armature phase voltages and currents are balanced three-phase quantities. Hence, if zero time is taken as the instant when phase A is on the direct-axis:

$$\omega = \omega_0,$$

$$\theta = \omega_0 t,$$

$$u_z = i_z = 0.$$

From the transformation equations (4.5) and (4.7), the voltage and current in phase A are:

$$\left.\begin{aligned} u_a &= u_d \cos \omega_0 t + u_q \sin \omega_0 t, \\ i_a &= i_d \cos \omega_0 t + i_q \sin \omega_0 t. \end{aligned}\right\} \qquad (7.1)$$

Now if the voltage and current in phase A are represented by phasors $\boldsymbol{U}$ and $\boldsymbol{I}$, as in Fig. 3.9, the components of the phasors are related to the axis quantities as shown below. The component phasors of current are indicated in Fig. 3.9 by the symbols $\boldsymbol{I}_d$ and $\boldsymbol{I}_q$ in heavy type. Let the magnitudes of the components be denoted by I_d and I_q in ordinary type. Then

$$\boldsymbol{I} = \boldsymbol{I}_d + \boldsymbol{I}_q = I_d + jI_q$$

Hence

$$\boldsymbol{I}_d = I_d + j \cdot 0,$$
$$\boldsymbol{I}_q = 0 + jI_q$$

Similarly let the magnitudes of the components of voltage be U_d and U_q.

$$\left.\begin{aligned} i_a &= \mathrm{Re}[\sqrt{2}(I_d + jI_q)\,\epsilon^{j\omega_0 t}] \\ &= \sqrt{2}(I_d \cos\omega_0 t - I_q \sin\omega_0 t), \\ u_a &= \sqrt{2}(U_d \cos\omega_0 t - U_q \sin\omega_0 t) \end{aligned}\right\} \qquad (7.2)$$

Since both pairs of relations stated by Eqns. (7.1) and (7.2) hold for all values of t, it follows, by equating coefficients, that

$$i_d = \sqrt{2}I_d, \qquad i_q = -\sqrt{2}I_q,$$
$$u_d = \sqrt{2}U_d, \qquad u_q = -\sqrt{2}U_q.$$

Hence for steady operation at synchronous speed the axis voltages and currents are all constant quantities independent of time. Moreover u_f and i_f are constant and $u_{kd}, u_{kq}, i_{kd}, i_{kq}$ are all zero. The general equations (4.27) can therefore be simplified, putting $p = 0$, and $\omega = \omega_0$ and using the synchronous reactances X_d, X_q, defined on p. 89.

$$\left.\begin{aligned} u_f &= R_f i_f \\ u_d &= R_a i_d + X_q i_q \\ u_q &= -X_{md} i_f - X_d i_d + R_a i_q \end{aligned}\right\} \qquad (7.3)$$

Hence

$$
\begin{aligned}
U &= U_{\mathrm{d}} + jU_{\mathrm{q}} \\
&= \frac{1}{\sqrt{2}}(u_{\mathrm{d}} - ju_{\mathrm{q}}) \\
&= \frac{1}{\sqrt{2}}[R_{\mathrm{a}}(i_{\mathrm{d}} - ji_{\mathrm{q}}) + X_{\mathrm{q}}i_{\mathrm{q}} + jX_{\mathrm{d}}i_{\mathrm{d}} + jX_{\mathrm{md}}i_{\mathrm{f}}] \\
&= R_{\mathrm{a}}(I_{\mathrm{d}} + jI_{\mathrm{q}}) + jX_{\mathrm{d}}I_{\mathrm{d}} - X_{\mathrm{q}}I_{\mathrm{q}} + \frac{jX_{\mathrm{md}}i_{\mathrm{f}}}{\sqrt{2}} \\
&= R_{\mathrm{a}}\boldsymbol{I} + jX_{\mathrm{d}}\boldsymbol{I}_{\mathrm{d}} + jX_{\mathrm{q}}\boldsymbol{I}_{\mathrm{q}} + \boldsymbol{U}_0,
\end{aligned}
\tag{7.4}
$$

where $U_0 = jX_{\mathrm{md}}i_{\mathrm{f}}/\sqrt{2}$, the no-load voltage. Eqn. (7.4) is the same as Eqn. (3.5). It is worth noting that the direction of rotation is here taken to be counterclockwise, whereas in Section 3.4 a clockwise rotation was chosen in order to simplify the derivation from the space phasor diagram of m.m.f. However the fundamental relationships are the same and lead to the same phasor equation.

The above analysis shows that for steady operation there is a direct relation between the axis values of voltage and current and the components of the phasors representing the phase values, and thus I_{d} and I_{q} are measurable quantities. The phasor diagram is in fact generally derived, as in Section 3.4, by using the phasor components. That method only applies, however, for a.c. conditions and cannot be used for general transient problems. When the phase voltages and currents do not vary sinusoidally with time the only way to replace them by axis quantities is to use the transformation Eqn. (4.4) to (4.7).

7.2 Starting of a synchronous motor

In this section the general equations of Chapter 4 are applied to an a.c. machine supplied with balanced three-phase voltages and running at a constant speed away from synchronism. This condition includes the normal steady operation of an induction motor and is also of considerable importance in analysing the behaviour of both induction motors and synchronous motors during starting. The analysis is based on Park's equations and applies directly to a synchronous machine with a damper winding on each axis in addition to the field winding. The results can be applied to the normal steady operation of an induction motor by treating it as a special case in which all the constants are the same for both axes.

In order to apply the analysis to the starting operation, it is necessary to make certain assumptions. The starting of a motor is initiated by switching on the supply when the motor is at rest. The starting operation is usually analysed by obtaining a torque-speed curve, which gives for each speed the torque that would be exerted if the motor ran steadily at that speed. Starting up is, however, a transient condition, during which the speed and all the other quantities are continually changing, and the instantaneous torque corresponding to any instantaneous speed actually differs to some extent from that determined in this way. If the rate of rise of speed is not too rapid the error is not great.

In an induction motor the speed rises smoothly until it settles down to the steady value just below synchronism. It is usually assumed that the torque exerted as the motor runs up is the same as that given by the torque-speed curve based on steady conditions. The method is accepted as a reliable way of studying the behaviour of an induction motor when starting up.

A synchronous motor, when switched on to the a.c. supply, runs up in a similar manner, with the damper and field windings carrying currents of slip frequency and acting like the secondary circuits of an induction motor. It cannot, however (apart from the effect of the small reluctance torque in a salient-pole machine), attain synchronous speed until a direct-current excitation in the field winding causes it to pull into step. The starting conditions are thus a good deal more complicated for the synchronous motor than for the induction motor. There are two possibilities:

1. If the field winding is connected to an exciter, which provides a voltage right from the time when the motor is at rest, the motor synchronizes without any further control operation.
2. If there is no exciter voltage in the field circuit during the running-up period, this excitation voltage must be brought up to a sufficient value before the motor can synchronize. The field winding is usually closed through a resistance during

starting, in order to limit the induced voltage when the motor is at stand-still. The synchronizing operation either switches in the exciter or, if the exciter armature has been included in the alternator field circuit, switches on the exciter field.

The starting of a synchronous motor must therefore be considered in two parts – running-up and synchronizing. The running-up period can be dealt with, in the same way as for the induction motor, by obtaining a torque-speed curve giving the torque that would be exerted if the motor ran at a constant speed. The solution to the synchronizing problem can only be found by solving the differential equations.

An induction motor running at a steady speed away from synchronism exerts a constant torque, but a synchronous motor, because of the lack of symmetry in the field system, develops a pulsating torque. Only the mean torque is of use in starting the motor, and consequently, in deriving the torque-speed curve, only the mean torque is considered.

The calculation assumes that the applied voltage u_f in the field circuit is zero. If u_f were not zero, there would be additional currents in the machine, which, by the principle of superposition, could be determined separately by assuming the armature voltage to be zero. The superimposed currents are those of a short-circuited generator operating at the appropriate frequency, and produce a braking torque. For this reason synchronous motors are usually started up with the field unexcited.

The aim of the following mathematical development is therefore to find how the axis currents and flux linkages vary as functions of time during the asynchronous period, assuming constant speed, and use this information in the torque expression of Eqn. (4.16) to find how the torque, calculated at a given speed, varies as a function of speed.

During the run-up period the rotor speed ω differs from the synchronous speed ω_0 of the air-gap m.m.f. by an amount defined as the slip speed $s\omega_0 = p\delta$. See Eqn. (5.1). If s is constant, the rotor angle increases at a uniform rate and, assuming the position of the rotating reference axis to correspond with the d-axis at zero time, $\delta = s\omega_0 t$. The expressions for the two axis components of armature voltage are found from Eqns. (5.4), (5.5), (4.19) and (4.22), with $u_f = 0$.

$$\left.\begin{aligned} U_m \sin s\omega_0 t &= p\psi_d + (1-s)\omega_0\psi_q + R_a i_d \\ U_m \cos s\omega_0 t &= -(1-s)\omega_0\psi_d + p\psi_q + R_a i_q \\ \bar{\psi}_d &= \frac{X_d(p)\bar{i}_d}{\omega_0} \\ \bar{\psi}_q &= \frac{X_q(p)\bar{i}_q}{\omega_0} \end{aligned}\right\} \quad (7.5)$$

Eqns. (7.5) are linear differential equations with sinusoidal applied voltages of slip frequency $s\omega_0/2\pi$. Hence there is a steady a.c. solution for which the axis quantities i_d, i_q, ψ_d, ψ_q all alternate at slip frequency. Thus when the motor runs at slip s, the phase currents alternate at supply frequency, but the axis currents alternate at slip frequency. The currents in the field and damper windings, which act like the secondary of an induction motor, also alternate at slip frequency.

The complex numbers representing the axis voltages can be deduced by putting u_d and u_q in the form:

$$u_d = \mathrm{Re}(-jU_m e^{js\omega_0 t}) = \mathrm{Re}(\sqrt{2}U_d e^{js\omega_0 t})$$
$$u_q = \mathrm{Re}(U_m e^{js\omega_0 t}) = \mathrm{Re}(\sqrt{2}U_q e^{js\omega_0 t})$$

Hence

$$\boldsymbol{U}_d = -jU_m/\sqrt{2} = -jU$$
$$\boldsymbol{U}_q = U_m/\sqrt{2} = U$$

where U is a scalar quantity whose magnitude is the r.m.s. value of the supply voltage. The phasor equations are obtained by substituting $p = js\omega_0$ in Eqns. (7.5) and replacing the small letters by capitals in heavy type:

$$\left.\begin{aligned} -jU &= js\omega_0\boldsymbol{\Psi}_d + (1-s)\omega_0\boldsymbol{\Psi}_q + R_a\boldsymbol{I}_d \\ U &= -(1-s)\omega_0\boldsymbol{\Psi}_d + js\omega_0\boldsymbol{\Psi}_q + R_a\boldsymbol{I}_q \\ \omega_0\boldsymbol{\Psi}_d &= X_d(js\omega_0)\boldsymbol{I}_d \\ \omega_0\boldsymbol{\Psi}_q &= X_q(js\omega_0)\boldsymbol{I}_q \end{aligned}\right\} \quad (7.6)$$

The mean torque M_e can be deduced from Eqn. (4.16) if the expression is modified so as to apply to the a.c. condition to yield

$$M_e = \frac{\omega_0}{2} \operatorname{Re}[\Psi_d^* I_q - \Psi_q^* I_d]$$

The mean motoring torque M_m is equal and opposite to M_e and is given by:

$$M_m = \frac{\omega_0}{2} \operatorname{Re}[\Psi_q^* I_d - \Psi_d^* I_q] \tag{7.7}$$

where Ψ_d^* and Ψ_q^* are the complex conjugates of Ψ_d and Ψ_q.

7.2.1 Torque-speed curve with primary resistance neglected

If R_a is neglected, the voltage equations (7.6) reduce to:

$$\left.\begin{aligned} -jU &= js\omega_0\Psi_d + (1-s)\omega_0\Psi_q, \\ U &= -(1-s)\omega_0\Psi_d + js\omega_0\Psi_q, \end{aligned}\right\} \tag{7.8}$$

whence

$$\begin{aligned} \omega_0\Psi_d &= -U \\ \omega_0\Psi_q &= -jU \end{aligned}$$

Substituting in Eqn. (7.7) and using Eqn. (7.6), the mean motoring torque becomes

$$\begin{aligned} M_m &= \frac{U^2}{2} \operatorname{Re}\left[\frac{1}{jX_d(js\omega_0)} + \frac{1}{jX_q(js\omega_0)}\right] \\ &= M_{m\,d} + M_{m\,q}, \end{aligned} \tag{7.9}$$

where

$$\left.\begin{aligned} M_{m\,d} &= \frac{U^2}{2} \operatorname{Re}\left[\frac{1}{Z_d}\right] \\ M_{m\,q} &= \frac{U^2}{2} \operatorname{Re}\left[\frac{1}{Z_q}\right] \end{aligned}\right\} \tag{7.10}$$

$$\left.\begin{aligned} Z_d &= \frac{1}{Y_d} = jX_d(js\omega_0) = j\frac{(1 + js\omega_0 T_d')(1 + js\omega_0 T_d'')}{(1 + js\omega_0 T_{d0}')(1 + js\omega_0 T_{d0}'')} X_d \\ Z_q &= \frac{1}{Y_q} = jX_q(js\omega_0) = j\frac{(1 + js\omega_0 T_q'')}{(1 + js\omega_0 T_{q0}'')} X_q \end{aligned}\right\} \tag{7.11}$$

using the expanded expressions for $X_d(p)$ and $X_q(p)$ stated in Eqns. (4.29) and (4.33). Y_d and Y_q are operational admittances, which are used in Eqn. (7.12).

The torque is thus the sum of two quantities, each of which is a torque associated with one axis only. A typical torque-speed curve of a synchronous motor, calculated by the above method, is shown on Fig. 7.1, where the thick line gives the total mean torque M_m and the thin lines give the two component torques M_{md} and M_{mq}.

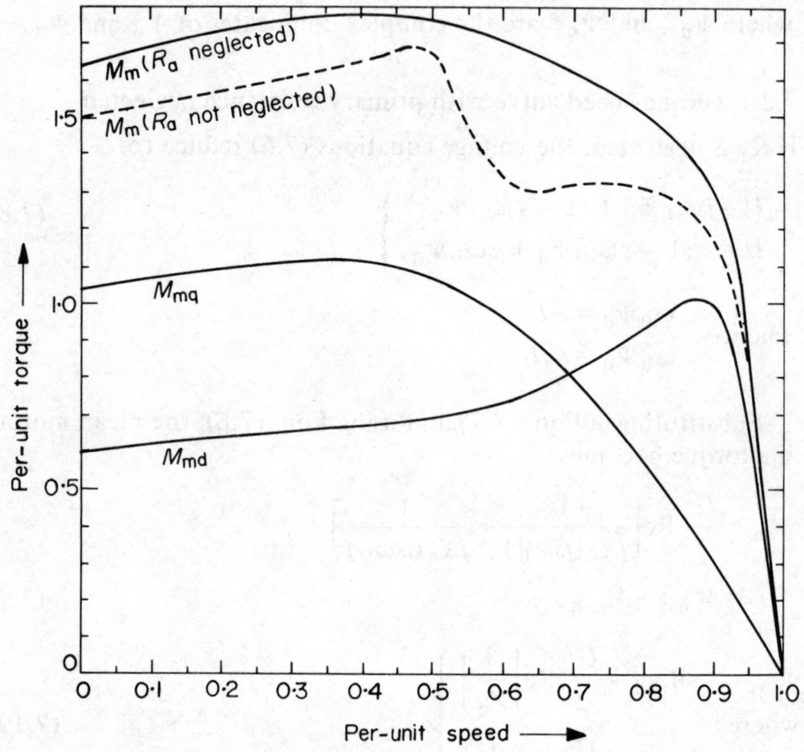

Fig. 7.1 Torque-speed curves of a synchronous motor.

7.2.2 Equivalent circuits

The operational equivalent circuits of Fig. 4.5 can be converted into a.c. equivalent circuits by substituting $p = js\omega_0$ and replacing small by capital symbols for the variables, as in the equations. If in addition the two applied voltages and all the impedances are divided by s, and u_f is put equal to zero, the equivalent circuits of

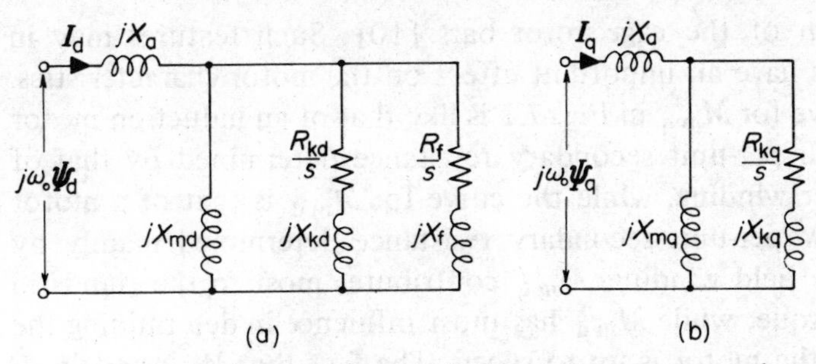

Fig. 7.2 Equivalent circuits for calculating the torque-speed curve of a synchronous motor.
(a) Direct axis (b) Quadrature axis

Fig. 7.2 are obtained. It is evident that the overall impedance of the direct-axis circuit (Fig. 7.2a) is $\boldsymbol{Z}_d$, since, from Eqns. (7.6):

$$j\omega_0\Psi_d = jX_d(js\omega_0)\boldsymbol{I}_d = \boldsymbol{Z}_d\boldsymbol{I}_d.$$

Similarly the impedance of the quadrature-axis circuit (Fig. 7.2b) is $\boldsymbol{Z}_q$.

In order to calculate the torque-speed curve of a synchronous motor, it is simpler to use the equivalent circuits instead of the formulas in Eqn. (7.9). The method is then similar to that commonly used for the induction motor, except that two calculations must be made. The quadrature-axis circuit (Fig. 7.2b) is similar to the equivalent circuit of the induction motor (Fig. 3.4) with the primary resistance neglected. Fig. 7.2a is similar to the equivalent circuit of a double-cage induction motor with two secondary circuits. Thus each of the component torques is calculated from the appropriate equivalent circuit in the same way as for an induction motor.

If the above results are applied to an induction motor, treated as a special case in which there is no field winding and all the constants for the direct and quadrature-axes are the same, the two equivalent circuits become identical. Hence only one circuit is needed, and the torque calculated from it is the required induction motor torque.

By using the equivalent circuit method, it is easy to obtain a more accurate determination of the torque of those synchronous machines for which the simplifications introduced in Section 4.4 would cause too much error. It is not difficult to set up more elaborate equivalent circuits, for example to allow for the distribution of the cage rotor bars [10]. Such features may in some cases have an important effect on the motor characteristics.

The curve for $M_{m\,q}$ in Fig. 7.1 is like that of an induction motor with a high per-unit secondary resistance determined by that of the damper winding, while the curve for $M_{m\,d}$ is that of a motor with a low per-unit secondary resistance determined mainly by that of the field winding. $M_{m\,q}$ contributes most to the standstill starting torque, while $M_{m\,d}$ has most influence in determining the slip when the motor is up to speed. The fact that $M_{m\,d}$ and $M_{m\,q}$ differ indicates that there is a large pulsating torque which is important in determining whether the motor can pull into synchronism when the field is applied.

7.2.3 Torque-speed curve allowing for primary resistance

The torque-speed curve calculated by neglecting R_a is accurate enough for many practical purposes. The curves obtained from tests of synchronous motors differ from the curves calculated by this method in having a double kink at about a half of synchronous speed, as shown dotted in Fig. 7.1. If the resistance is not neglected, the theoretical curve obtained by a more exact calculation shows a kink of this type. There is also an overall loss of torque when the resistance is taken into account.

Eqns. (7.6) can be rewritten, using Eqns. (7.11):

$$\left.\begin{aligned} -jU &= j(s + R_a Y_d)\omega_0 \Psi_d + (1-s)\omega_0 \Psi_q \\ U &= -(1-s)\omega_0 \Psi_d + j(s + R_a Y_q)\omega_0 \Psi_q \end{aligned}\right\} \tag{7.12}$$

From Eqns. (7.7) and (7.12), the following expression is obtained for the mean torque:

$$\left.\begin{aligned} M_m = \frac{U^2}{2|D|^2} \cdot \text{Re}\Big[&(Y_d + Y_q) + \frac{R_a}{\alpha}(2Y_d Y_q + |Y_d|^2 + |Y_q|^2) \\ &+ \frac{R_a{}^2}{\alpha^2}(Y_q |Y_d|^2 + Y_d |Y_q|^2)\Big] \end{aligned}\right\} \tag{7.13}$$

where $D = 1 - s\dfrac{R_a}{\alpha}(Y_d + Y_q) - \dfrac{R_a{}^2}{\alpha} Y_d Y_q$,

$\alpha = 1 - 2s.$

The dotted curve of Fig. 7.1 was calculated from Eqn. (7.13). The presence of the factor $\alpha = (1 - 2s)$ in the denominator of some of the terms explains why the effect of R_a is greater near to half speed, when the value of α is small [17].

7.3 Negative-sequence reactance of a synchronous machine

In the application of symmetrical component theory to power-system analysis, a synchronous generator is assumed to have a definite negative-sequence reactance. The validity of this assumption and the determination of the appropriate value can be studied by the method of Section 7.2.

7.3.1 Current due to negative sequence voltage

In a machine running at synchronous speed with negative-sequence applied voltages, the current is the same as it would be if the machine ran at a negative speed equal to synchronous speed and had positive-sequence applied voltages. Hence the theory of the last section applies with $s = 2$. With such a large slip the factors of the form $(1 + 2j\omega_0 T)$ in the expressions for Z_d and Z_q in Eqns. (7.11) can all be replaced approximately by $2j\omega_0 T$, since $2\omega_0 T$ is much greater than 1. Hence, if R_a is neglected, Eqns. (7.6) and (7.8) give:

$$I_d = \frac{-U}{X_d(2j\omega_0)} = \frac{-U}{X_d} \cdot \frac{T_{d0}'T_{d0}''}{T_d'T_d''} = -\frac{U}{X_d''} \text{ (approx.)},$$

$$I_q = \frac{-jU}{X_q(2j\omega_0)} = \frac{-jU}{X_q} \cdot \frac{T_{q0}''}{T_q''} = \frac{-jU}{X_q''}$$

In order to find the phase current i_a the instantaneous axis currents i_d and i_q are determined as follows, putting $U = U_m/\sqrt{2}$:

$$\left.\begin{aligned} i_d &= \mathrm{Re}(\sqrt{2} I_d \epsilon^{2j\omega_0 t}) = \frac{-U_m}{X_d''} \cos 2\omega_0 t, \\ i_q &= \mathrm{Re}(\sqrt{2} I_q \epsilon^{2j\omega_0 t}) = \frac{U_m}{X_q''} \sin 2\omega_0 t. \end{aligned}\right\} \quad (7.14)$$

Using the transformation Eqn. (4.5) with $\theta = (1-s)\omega_0 t$

$$i_a = -U_m\left[\frac{1}{X_d''}\cdot\cos 2\omega_0 t\cdot\cos\omega_0 t + \frac{1}{X_q''}\sin 2\omega_0 t\cdot\sin\omega_0 t\right]$$

$$= -\frac{U_m}{2}\left[\frac{1}{X_d''}+\frac{1}{X_q''}\right]\cos\omega_0 t - \frac{U_m}{2}\left[\frac{1}{X_d''}-\frac{1}{X_q''}\right]\cos 3\omega_0 t \qquad (7.15)$$

The applied voltage is:

$$u_a = U_m \sin\omega_0 t$$

Thus the current due to the negative-sequence applied voltage contains a third harmonic in addition to the fundamental. If the fundamental only is considered, the effective negative-sequence reactance is:

$$X_{2u} = \frac{2X_d''X_q''}{X_d''+X_q''} \qquad (7.16)$$

7.3.2 Voltage due to negative-sequence current

The voltage due to balanced negative-sequence currents can be found by considering a machine running with slip $s = 2$, and carrying positive-sequence currents. If the same assumptions are now made for the current that were previously made for the voltage:

$$i_a = I_m \sin\omega_0 t$$

$$I_d = 0 - \frac{jI_m}{\sqrt{2}}$$

$$I_q = \frac{I_m}{\sqrt{2}} + j0$$

From Eqns. (7.6) with $s = 2$ and R_a neglected:

$$U_d = 2j\omega_0\Psi_d - \omega_0\Psi_q,$$

$$U_q = \omega_0\Psi_d + 2j\omega_0\Psi_q,$$

where, as before,

$$\omega_0\Psi_d = X_d''I_d$$

$$\omega_0\Psi_q = X_d''I_q$$

Hence

$$U_d = (2X_d'' - X_q'')\frac{I_m}{\sqrt{2}}$$

$$U_q = (X_d'' - 2X_q'')\frac{-jI_m}{\sqrt{2}}$$

$$u_d = I_m\,(2X_d'' - X_q'')\cos 2\omega_0 t$$

$$u_q = -I_m\,(2X_q'' - X_d'')\sin 2\omega_0 t$$

whence

$$u_a = -I_m\,[(2X_d'' - X_q'')\cos 2\omega_0 t \cdot \cos \omega_0 t - (2X_q'' - X_d'')\sin 2\omega_0 t \cdot \sin \omega_0 t]$$

$$= -\frac{I_m}{2}(X_d'' + X_q'')\cos \omega_0 t + \frac{3I_m}{2}(X_d'' - X_q'')\cos 3\omega_0 t \quad (7.17)$$

Hence with balanced negative-sequence currents the voltage contains a third harmonic. The value of negative-sequence reactance based on the fundamental voltage is:

$$X_{2i} = \tfrac{1}{2}(X_d'' + X_q'') \quad (7.18)$$

7.3.3 Value of negative-sequence reactance

For the more general case of unbalanced conditions, when the external load is unbalanced, the appropriate value of negative-sequence reactance, defined as the ratio of fundamental voltage to fundamental current, is not necessarily either of the values determined above. If X_d'' and X_q'' are nearly equal, as they should be in a good design, the range of possible values is small and the associated third harmonic components are also small. The matter is discussed further in Section 8.4.

7.4 Small changes relative to a steady state

An important class of problems is that in which a machine, while operating at a steady state, is subjected to a small disturbance. The theory is developed by *linearizing* the equations according to the method explained in Section 2.3. The method can be applied to a small change of any kind, but in most practical problems the superimposed change oscillates sinusoidally at a given frequency. A distinction can be made between *forced oscillations* and *free oscillations*.

Forced oscillations occur when there is an externally applied torque or voltage, like the low-frequency pulsation set up in a diesel engine driving a generator, which causes all the quantities in the system to pulsate at the same frequency. The problem is to determine the values of any mechanical stresses or voltage fluctuations which may cause difficulties in operation.

Free oscillations occur without any externally applied impulse. They arise by a process of self-excitation either in the machine itself or as a result of closed-loop feedback circuits around the machine, like a voltage regulator or a speed governor. To calculate the behaviour of the system, it is necessary to imagine a small disturbance and to determine whether the oscillation is sustained after the disturbance is removed.

The theoretical treatment must first establish the values of all variables in the basic steady state, which then appear in some of the coefficients of the linearized equations. The following equations apply to both forced and free oscillations and do not allow for saturation.

7.4.1 Equations for steady operation

The equations for steady operation are obtained by putting $p = 0$ in the six Eqns. (4.16), (4.19), (4.22), (5.1), (5.2) and (5.4). The steady values, about which small changes occur, are indicated by the suffix 0 and complex numbers representing small oscillations are denoted by a bold capital letter with the prefix Δ; for example, the pulsation of the direct axis current is $\Delta \mathbf{I}_d$.

$$\left.\begin{aligned} U_m \sin \delta_0 &= \omega_0 \psi_{q0} + R_a i_{d0} \\ U_m \cos \delta_0 &= -\omega_0 \psi_{d0} + R_a i_{q0} \end{aligned}\right\} \qquad (7.19)$$

$$\left.\begin{aligned} \omega_0 \psi_{d0} &= X_d i_{d0} + X_{md} u_{f0}/R_f \\ \omega_0 \psi_{q0} &= X_q i_{q0} \end{aligned}\right\} \qquad (7.20)$$

$$M_0 = -M_{e0} = \frac{\omega_0}{2} [\psi_{d0} i_{q0} - \psi_{q0} i_{d0}] \qquad (7.21)$$

A negative sign appears in Eqn. (7.21) because the externally applied torque is positive when the machine acts as a motor.

7.4.2 Equations for small oscillations in the absence of regulators

The equations for small oscillations at frequency $m/2\pi$ are derived from the same six equations by the method explained in Section 2.3, and putting $p = jm$. Since the field voltage is constant,

$$\Delta U_f = 0$$

$$U_m \cos \delta_0 \cdot \Delta\delta = jm \cdot \Delta\Psi_d + \omega_0 \cdot \Delta\Psi_q + R_a \cdot \Delta I_d - jm\psi_{q0} \cdot \Delta\delta$$

$$-U_m \sin \delta_0 \cdot \Delta\delta = -\omega_0 \cdot \Delta\Psi_d + jm \cdot \Delta\Psi_q + R_a \cdot \Delta I_q + jm\psi_{d0}\, \Delta\delta \tag{7.22}$$

$$\left.\begin{aligned} \omega_0 \cdot \Delta\Psi_d &= X_d(jm) \cdot \Delta I_d \\ \omega_0 \cdot \Delta\Psi_q &= X_q(jm) \cdot \Delta I_q \end{aligned}\right\} \tag{7.23}$$

$$\Delta M = -\Delta M_e = \frac{-\omega_0}{2}[i_{q0} \cdot \Delta\Psi_d + \psi_{d0} \cdot \Delta I_q - i_{d0} \cdot \Delta\Psi_q - \psi_{q0} \cdot \Delta I_d] \tag{7.24}$$

7.5 Approximate methods for forced oscillations

The problem already mentioned of a diesel-driven synchronous generator connected to an infinite bus, provides a good example of a forced oscillation. Simple approximate formulas, which give sufficient accuracy for the calculation of forced oscillations, can be derived by neglecting the armature resistance R_a.

With a known field voltage u_{f0}, let U and U_0 be the respective r.m.s. values of the applied voltage and of the voltage induced on open circuit by u_{f0}. Then

$$U = \frac{U_m}{\sqrt{2}}$$

$$U_0 = \frac{1}{\sqrt{2}}\left[\frac{-X_{md}u_{f0}}{R_f}\right]$$

The negative sign in the expression for U_0 indicates that, with the sign conventions used, a negative field voltage is required to generate a positive armature voltage. Substituting these values in

Eqns. (7.19) and (7.20), and putting $R_a = 0$, the following expressions for the steady currents and flux linkages are obtained.

$$\left.\begin{aligned} \omega_0 \psi_{d0} &= -\sqrt{2}U \cos \delta_0 \\ \omega_0 \psi_{q0} &= \sqrt{2}U \sin \delta_0 \\ i_{d0} &= \frac{\sqrt{2}}{X_d}(-U \cos \delta_0 + U_0) \\ i_{q0} &= \frac{\sqrt{2}}{X_q} U \sin \delta_0 \end{aligned}\right\} \tag{7.25}$$

Substituting ψ_{d0} and ψ_{q0} from Eqns. (7.25) in Eqns. (7.22), with $R_a = 0$, and rearranging, gives the following pair of simultaneous equations:

$$jm \cdot \Delta\Psi_d + \omega_0 \cdot \Delta\Psi_q = \frac{\sqrt{2}U}{\omega_0}(jm \sin \delta_0 + \omega_0 \cos \delta_0) \cdot \Delta\delta,$$

$$-\omega_0 \cdot \Delta\Psi_d + jm \cdot \Delta\Psi_q = \frac{\sqrt{2}U}{\omega_0}(-\omega_0 \sin \delta_0 + jm \cos \delta_0) \cdot \Delta\delta.$$

The solution of these equations, in conjunction with Eqns. (7.23) gives the oscillations of the currents and flux linkages.

$$\left.\begin{aligned} \omega_0 \cdot \Delta\Psi_d &= \sqrt{2}U \sin \delta_0 \cdot \Delta\delta \\ \omega_0 \cdot \Delta\Psi_q &= \sqrt{2}U \cos \delta_0 \cdot \Delta\delta \\ \Delta I_d &= \frac{\sqrt{2}U}{X_d(jm)} \sin \delta_0 \cdot \Delta\delta \\ \Delta I_q &= \frac{\sqrt{2}U}{X_q(jm)} \cos \delta_0 \cdot \Delta\delta \end{aligned}\right\} \tag{7.26}$$

Hence substituting Eqns. (7.25) and (7.26) in Eqn. (7.24):

$$\begin{aligned} \frac{\Delta M}{\Delta\delta} &= \frac{UU_0}{X_d} \cos \delta_0 + U^2 \cos^2 \delta_0 \cdot \left[\frac{1}{X_q(jm)} - \frac{1}{X_d}\right] \\ &\qquad + U^2 \sin^2 \delta_0 \cdot \left[\frac{1}{X_d(jm)} - \frac{1}{X_q}\right] \\ &= K + jmC \end{aligned} \tag{7.27}$$

The parameters K and C, obtained by calculating the real and imaginary parts of Eqn. (7.27), are analogous to the elastic and damping constants of the linear, two-dimensional, mechanical system of Fig. 5.1, for which the equation of motion would be

$$-m^2 J \cdot \Delta\delta + jmC \cdot \Delta\delta + K \cdot \Delta\delta = \Delta M_t \tag{7.28}$$

The angular pulsation of the rotor due to a torque pulsation ΔM_t is

$$\Delta\delta = \frac{\Delta M_t}{(K - m^2 J) + jmC} \tag{7.29}$$

However the parameters K and C of the electro-mechanical system are functions of m and are far from constant. The calculated values for a 2600 kVA, 28-pole, diesel-driven generator are given in Fig. 7.3.

An earlier method of calculating the forced oscillations of a diesel-driven generator was based on the rather extreme assumption that the relation between torque and rotor angle could be taken as the slope of the steady-state power-angle curve (Fig. 3.11). It is readily shown that the synchronizing torque coefficient P_0 in Eqn. (3.9) is equal to the value of K at zero frequency. Also the damping factor C was assumed to be zero, which again is the zero-frequency value of the more exact calculation. These assumptions led to the concept of an electromagnetic *natural frequency* of a synchronous machine.

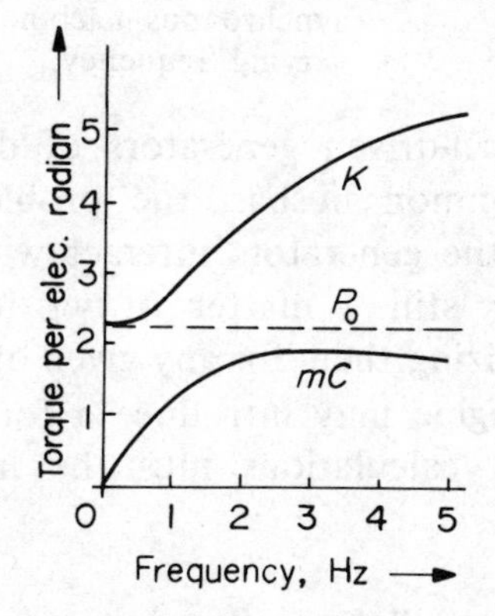

Fig. 7.3 Elastic and damping constants of a synchronous machine at varying frequency.

If there were no damping, Eqn. (7.29) shows that for a given torque pulsation, the angular deviation would become infinite when $K = m^2 J$. The frequency at which this occurs is the natural frequency and is given by

$$\frac{m_0}{2\pi} = \frac{1}{2\pi}\sqrt{\frac{K}{J}} \tag{7.30}$$

Fig. 7.4 gives two response curves for the diesel-driven generator. It shows the value of $\Delta\delta$ for a fixed amplitude of torque pulsations at different frequencies; the natural frequency occurs at about 2 Hz. Clearly the earlier method is inaccurate when there is a damper winding which causes K to increase and C to have an appreciable value at the frequency of the forced oscillation.

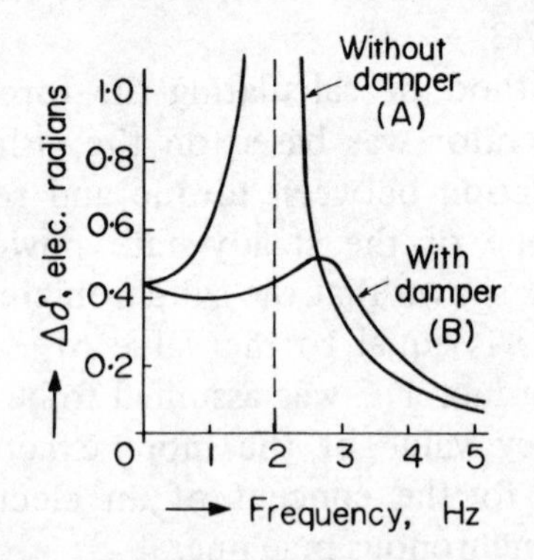

Fig. 7.4 Angular pulsation of a synchronous machine at varying frequency.

When several diesel-driven generators of different design are connected to a common busbar, the problem becomes more complicated, since the generators interact with each other. The analysis is however still a matter of writing the full set of equations and linearizing them for any given operating conditions. Since each diesel engine may introduce a forced oscillation of a different frequency, calculations must be made for each frequency.

7.6 Free oscillations. Steady-state stability

Free oscillations, otherwise known as *hunting*, *self-excitation* or *instability*, arise when the connections of the machine or system are such that the oscillation is sustained without any external agency. In a completely linear system, which fortunately cannot exist, the oscillation would rise to infinity. In a practical system the non-linearities limit the sustained oscillation to a value which cannot be regarded as 'small'.

Free oscillations often occur in a small synchronous machine, particularly if it has no damper winding, and they can be quite troublesome. It is readily shown from Eqn. (7.27) that, if R_a is zero, the damping constant is always positive, and hence that self-excitation is not possible. The criterion for instability is that the damping parameter C must be negative at some value of frequency. It depends very much on the presence of the armature resistance R_a. The following formula for determining K and C, for a machine connected to an infinite bus, is derived from Eqns. (7.22) to (7.24) without neglecting R_a. It is obtained by assuming that m is small compared with ω, so that $\omega^2 - m^2 \approx \omega^2$:

$$\frac{\Delta M}{\Delta \delta} = \frac{1}{2D}\Big\{\omega_0{}^2(A\psi_{q0}{}^2 + B\psi_{d0}{}^2) - \omega_0(\psi_{d0}i_{d0} + \psi_{q0}i_{q0})$$

$$+\frac{jm}{\omega_0}R_a[\omega_0{}^2AB(\psi_{d0}{}^2 + \psi_{q0}{}^2) - \omega_0(B-A)(\psi_{q0}i_{q0} - \psi_{d0}i_{d0})$$

$$-(i_{d0}{}^2 + i_{q0}{}^2)]$$

$$+R_a{}^2[\omega_0\delta AB(\psi_{d0}i_{d0} + \psi_{q0}i_{q0}) - (Ai_{d0}{}^2 + Bi_{q0}{}^2)]\Big\}$$

$$= K + jmC \tag{7.31}$$

Where

$$A = \frac{1}{X_d(jm)}$$

$$B = \frac{1}{X_q(jm)}$$

$$D = 1 + \frac{jm}{\omega_0}R_a(A+B) + R_a{}^2AB.$$

In a large power system, to which are connected many generators and motors, the problem of stability is of great importance. The steady-state stability of a system determines whether the system is able to operate stabily at any steady condition. It can be studied by deriving the linear equations for small changes or small oscillations, as in Sections 7.4 and 7.5, and determining whether or not the change dies away in time, thus restoring the system to its original state. Because of the non-linearity, the coefficients in the linearized equations are different for every operating point, with the result that the system is stable at some points and unstable at others. On the operating chart of Fig. 3.12, which applies to a single machine connected to an infinite bus, there is a stable region on the right and an unstable region on the left, separated by the stability limit curve MD. The position of the limit curve can be modified by the use of an excitation regulator or a turbine governor.

The stability of a generator without any regulating devices can be determined, as explained in standard text-books, by considering the steady power-angle curve of Fig. 3.11. At a point where the power increases with the angle, an increase of angle produces a change of torque which causes the machine to decelerate back to the original condition, and the operation is stable. At an angle beyond the maximum point of the curve, the machine continuously moves away from the original condition and is unstable. The instability is of the *aperiodic* or *drifting* type in which the speed drifts slowly away from the steady condition.

7.6.1 Equations for small changes when regulators are present

If regulating devices are present, it is not possible to use a simple criterion like the one just described. The regulators and machine form a closed-loop control system and it is necessary to set up linearized equations, which allow for all the regulators as well as the machine, before any of the available stability criteria can be applied. The instability may be of the *drifting* type, or it may be *oscillatory* depending on the gain of the regulator.

In addition to the assumption made in Section 7.4, it is assumed for the purpose of this section that, since the speed change during a small disturbance period is negligible, the voltage terms $\psi_q\dot{\delta}$, $\psi_d\dot{\delta}$, $\dot{\psi}_d$ and $\dot{\psi}_q$ in Eqn. (5.5) are negligible compared with the voltages generated by the fluxes rotating at synchronous speed.

From a general point of view a machine, which has a quadrature field winding of the type described in Section 6.3, has three control inputs and three control outputs. The equations are those on pp. 87 and 88 with the modifications explained on p. 129 including the introduction of a quadrature transfer function $G_q(p)$. The linearized machine equations can be expressed in the following form in the Laplace domain [47]:

$$\begin{bmatrix} G_d(p) \cdot \overline{\Delta u_{fd}} \\ \overline{\Delta M_t} \\ G_q(p) \cdot \overline{\Delta u_{fq}} \end{bmatrix} = \begin{bmatrix} -X_d(p) & U_m \sin \delta_0 & -R_a \\ \dfrac{-U_m \sin \delta_0}{2} + R_a i_{d0} & -(Q_0 + Jp^2) & \dfrac{-U_m \cos \delta_0}{2} + R_a i_{q0} \\ -R_a & U_m \cos \delta_0 & -X_q(p) \end{bmatrix} \begin{bmatrix} \overline{\Delta i_d} \\ \overline{\Delta \delta} \\ \overline{\Delta i_q} \end{bmatrix} \quad (7.32)$$

Where

$$\left.\begin{aligned} G_d(p) &= \frac{X_{md}(1 + T_{kd}p)}{R_{fd}(1 + T_{d0}'p)(1 + T_{d0}''p)} \\ G_q(p) &= \frac{X_{mq}(1 + T_{kq}p)}{R_{fq}(1 + T_{q0}'p)(1 + T_{q0}''p)} \end{aligned}\right\} \quad (7.33)$$

and

$$Q_0 = \frac{U_m}{\sqrt{2}}(I_{d0} \cos \delta_0 + I_{q0} \sin \delta_0) \quad (7.34)$$

is the reactive power. For a machine with a single field winding the equations have the same form, but $G_q(p) = 0$.

The three control inputs are the two field voltages u_{fd} and u_{fq} and the turbine torque M_t. The three control outputs are the two axis currents i_d and i_q and the rotor angle δ. Any feedback signal injected into any kind of excitation regulator or turbine governor can be expressed as a function of the three control outputs.

The matrix in Eqn. (7.32) may be inverted to yield the

machine's transfer function which is then combined with the regulator transfer function to find the required overall loop transfer function and the characteristic equation of the control system. Any of the stability criteria such as the Nyquist-, Routh-, or root-locus methods described in textbooks on control theory, can be applied to the loop transfer function.

The results of two studies, in which micro-machine measurements were compared with theoretical computations, are described below. The first applies to a generator with the commonly used voltage regulator in which a terminal voltage signal, modified by the regulator transfer function, is fed back to a conventional direct-axis field winding. The second is a system like the one shown in Fig. 6.3, where, in addition to the conventional voltage regulator, a special angle regulator feeds back a signal depending on the rotor angle to a quadrature-axis field winding.

The effect of the regulators is illustrated by the $P-Q$ chart of Fig. 7.5 which compares the stability limit curves for generators with fixed excitation, with a normal voltage regulator, and with a quadrature-axis regulator. Since a margin of stability must be allowed, operation with leading power factor is limited for a generator with fixed excitation. A voltage regulator on a con-

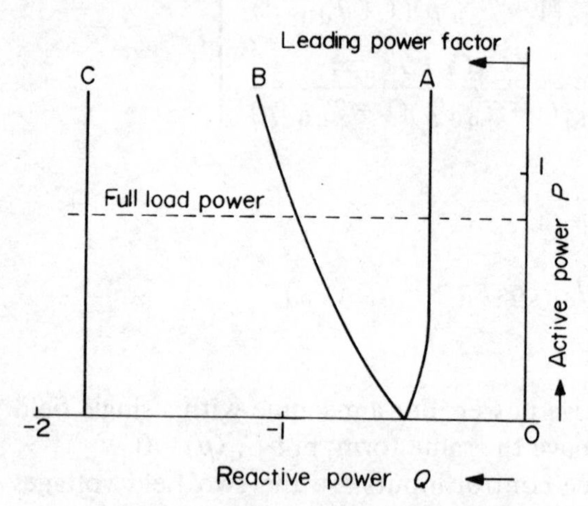

Fig. 7.5 *P-Q* chart showing stability limit curves with different types of excitation.
A – Fixed excitation.
B – Direct-axis regulation.
C – Quadrature-axis regulation.

ventional machine extends the range of operation only when the generator is loaded. The quadrature-axis excitation regulation on the other hand allows generation at leading power factor at all loads, the current being limited only by heating.

7.6.2 Voltage regulator acting on a direct-axis field

Different types of regulators have been described in Section 6.2 [33], but for the purpose of this illustration a proportionate regulator only is considered. The generator is connected to an infinite bus through a fixed-impedance as shown in Fig. 6.2 and the turbine torque is assumed to be constant. Equation (7.32) is used to represent the generator but ΔM_t and Δu_{fq} are equal to zero and the only control input is the field voltage Δu_{fd}.

A test or calculation is made at a given operating point on the P–Q chart to determine whether the system is stable or unstable at that point. The theoretical stability limit is calculated by studying for example the loci of the three roots of the characteristic equation in the complex plane of Fig. 7.6. The curves in Fig. 7.6a show that, without a regulator, an increase in the steady-state value of rotor angle δ_0 causes the complex roots to move further into the left-half plane but the real root moves through the origin into the right-half plane. For the numerical example under consideration this occurs when $\delta_0 = 84°$. Once positive, the real root contributes to an exponential term with a positive exponent in the time domain and causes instability of the drifting type.

The root-loci in Fig. 7.6b show the effect of regulator gain K at a given value of δ_0 equal to 100°. With zero gain the position of the roots are identical to those in Fig. 7.6a at $\delta_0 = 100°$ and the system drifts out of stability. However, as the value of K is increased, the real root moves to the left in Fig. 7.6b and crosses into the left-half plane when $K = 2.2$, but at the same time the complex roots move to the right and when $K = 16$ they cross the imaginary axis and give rise to oscillations. Hence at $\delta_0 = 100°$ the system is stable provided

$$2.2 < K < 16 \tag{7.35}$$

Root-loci like those in Fig. 7.6b, but for different values of δ_0, at constant P, have been studied to obtain a series of results like

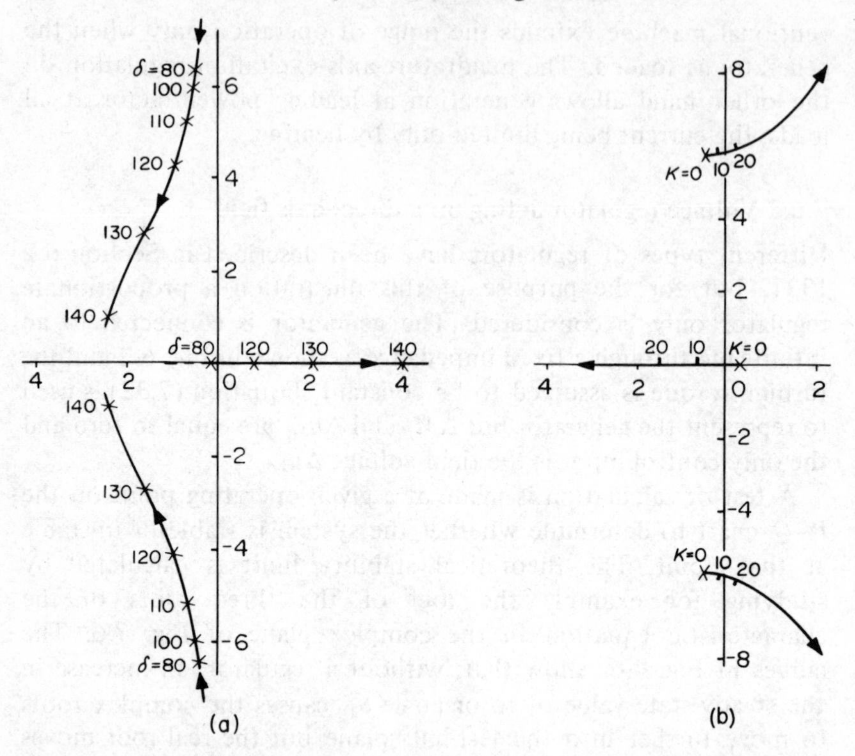

Fig. 7.6 **(a)** Unregulated system characteristic equation roots for different values of rotor angle δ.
(b) Root-loci for the system with voltage feedback at δ = 100 degrees.

Eqn. (7.35). Fig. 7.7 shows how δ at the stability limit varies with regulator gain. The left hand branch in Fig. 7.7 defines the boundary where drifting instability sets in and the right hand branch defines the boundary of oscillatory instability.

7.6.3 Controlled quadrature excitation

As perviously mentioned, the range of stable operation of a generator connected to an infinite bus through a fixed reactance, can be extended still further by using a special regulator to control the excitation voltage applied to a quadrature field winding on the generator. The most successful feedback signal for the quadrature regulator is one derived from the generator rotor angle [44]. The angle regulator holds the rotor angle, and hence the rotor position with respect to a synchronously rotating reference frame (see p.

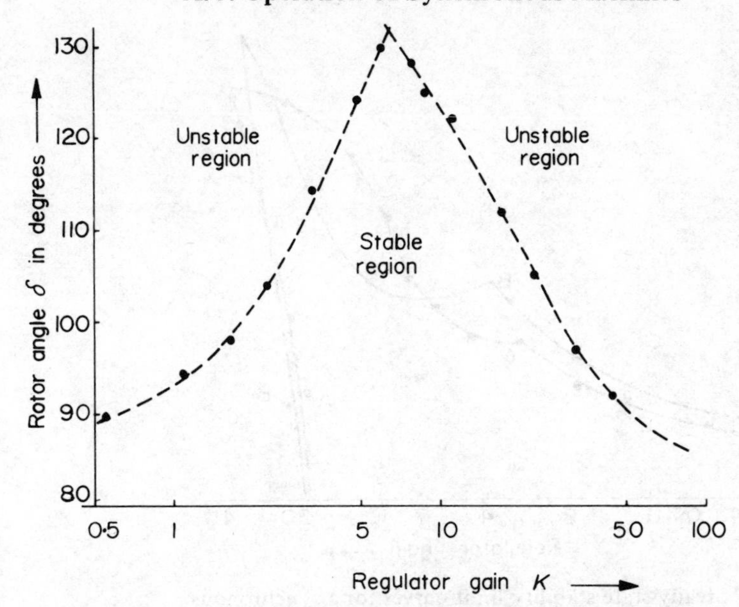

Fig. 7.7 Measured steady-state stability limit curve for a conventional synchronous machine when the terminal voltage is used to regulate the direct-axis field excitation. The active power is 0.8 p.u.

100), nearly constant and stability is thereby improved because the flux position is adjusted electrically in space instead of requiring mechanical movement of the rotor.

Stability tests on a micro-machine were first made with a constant excitation voltage applied to the direct-axis field and an angle regulator, with proportionate, first- and second derivative terms, applied to the quadrature-axis field. It was found that, with a suitable value of regulator gain, stability could be maintained at all levels of active power in the leading region up to a negative reactive power of about 3 per-unit. In Fig. 7.8 the discrepancy between curves A and B is thought to be due to saturation of the leakage flux paths.

Further stability tests on a micro-machine with a divided winding (Fig. 6.4) are described in [47]. They were carried out with the voltage regulator and the angle regulator both in operation while maintaining constant prime mover torque. The results show the voltage regulator has less effect on the stability than the angle regulator, in fact, the angle feedback stabilizes the

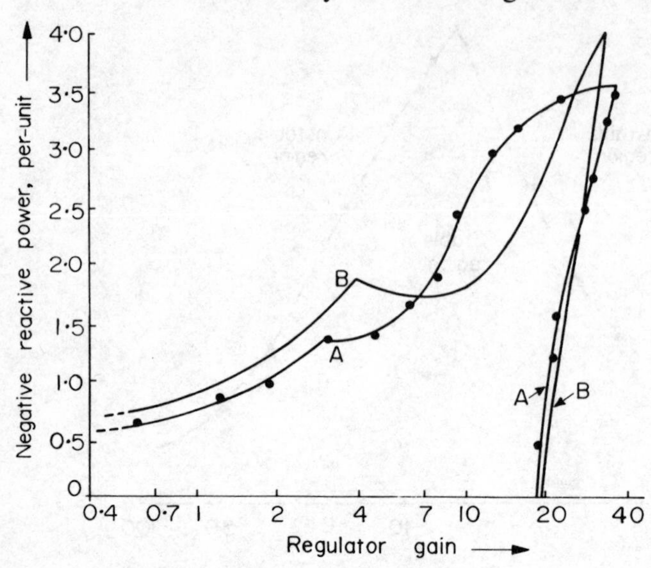

Fig. 7.8 Steady-state stability limit curves for a synchronous machine, controlled as in Fig. 6.3, but with fixed direct-axis excitation.
A – test curve; B – theoretical, neglecting saturation.

voltage feedback and permits the use of high values of voltage regulator gain.

The stability calculations for both regulators in operation were based on Eqn. (7.32) with ΔM_t equal to zero but with two control inputs viz. Δu_{fd} and Δu_{fq}. The transformation Eqn. (6.5) was used. One of the two feedback regulator loops, with a given value of regulator gain K_1, is regarded as part of the forward loop of the control system, and the remaining regulator loop with gain K_2 is regarded as open. The Nyquist, root-locus or other stability criteria may be applied to the overall open-loop transfer function to find the upper and lower stable limits of K_2 for different operating points on the P–Q chart. The result is summarized by a stability limit curve, like those in Fig. 7.8, showing the stable values of negative reactive power as a function of K_2. However, such a curve is only valid for one value of K_1 and the calculation has to be repeated, giving a different stability limit curve for any other value of K_1.

As the number of feedback loops increase to say n, the above classical technique of regarding $(n-1)$ of the feedback loops as part of the forward loop, becomes more and more cumbersome although it allows some physical insight into the problem. The modern approach to study such a multi-input–output system is to use the state variable technique to establish the system matrix and the characteristic equation (see p. 111). For stable operation the eigen values or roots of the characteristic equation must have negative real parts. Fast digital computers using numerical matrix inversion methods make the eigen value approach particularly attractive for higher order practical systems.

Chapter Eight
Synchronous Generator Short-Circuit and System Faults

8.1 Symmetrical short-circuit of an unloaded synchronous generator

Although the majority of the faults occurring in practice on a power system are unsymmetrical between the phases, the symmetrical fault is important because, although rarer, it is more severe. It is, moreover, a simpler condition to analyse, and therefore forms a suitable starting-point for a study of fault conditions. The short-circuit test, in which the three terminals of an unloaded synchronous generator are all short-circuited simultaneously, is a well-established method of checking its transient characteristics. The present section gives a full analysis of a sudden symmetrical short-circuit of an unloaded synchronous generator, and the solution is then extended in Section 8.3 so that it can be applied to a loaded machine [24].

If the speed is not assumed constant during the transient, the problem belongs to the fifth category in Table 5.1 on p. 99 and a step-by-step method is needed to solve the non-linear equations as discussed in Section 9.2. However, in the present chapter the speed is assumed constant and the problem belongs to the third category in Table 5.1. The equations are linear and are solved by means of the principle of superposition and Laplace transforms; the notation used for the original steady quantities and for the superimposed values is that explained on p. 39.

8.1.1 The operational equations

Let $t = 0$ at the instant of short-circuit, and let λ be the angle between the axis of phase A and the direct-axis at that instant. The angle λ therefore defines the point in the a.c. cycle at which the short-circuit occurs. Then, assuming that the speed has the constant value ω_0:

$$\theta = \omega_0 t + \lambda$$

The equations of p. 134 hold for the steady values except that $\omega_0 t$ is replaced by $(\omega_0 t + \lambda)$ so that

$$u_{a0} = \sqrt{2}[U_{d0} \cos(\omega_0 t + \lambda) - U_{q0} \sin(\omega_0 t + \lambda)] \tag{8.1}$$

Comparing this with the transformation equation

$$u_{a0} = u_{d0} \cos(\omega_0 t + \lambda) + u_{q0} \sin(\omega_0 t + \lambda)$$

it follows that

$$u_{d0} = \sqrt{2} U_{d0} = U_{md}$$
$$u_{q0} = -\sqrt{2} U_{q0} = U_{mq}$$

In general the instantaneous values of the two components of u_{a0} are therefore

$$\left.\begin{aligned} (u_{a0})_d &= U_{md} \cos(\omega_0 t + \lambda) \\ (u_{a0})_q &= U_{mq} \sin(\omega_0 t + \lambda) \end{aligned}\right\} \tag{8.2}$$

However, for an unloaded machine, i_{d0} and i_{q0} are both zero, and hence, from Eqn. (7.3):

$$u_{d0} = U_{md} = 0$$
$$u_{q0} = U_{mq} = -X_{md} i_f = U_m = \sqrt{2} U$$

so that the original phase voltage becomes, from Eqn. (8.1),

$$u_{a0} = U_m \sin(\omega_0 t + \lambda)$$

The equations for the superimposed armature quantities are found from Eqn. (4.15) as

$$\left.\begin{aligned} u_d' &= \dot{\psi}_d' + \omega_0 \psi_q' + R_a i_d' \\ u_q' &= -\omega_0 \psi_d' + \dot{\psi}_q' + R_a i_q' \end{aligned}\right\} \tag{8.3}$$

When the fault occurs, the values of u_d and u_f do not change, so that $u_d' = u_f' = 0$, but the value of u_q changes abruptly from U_m

to zero. After the fault, i_d and i_q have the same values as i_d' and i_q' since $i_{d0} = i_{q0} = 0$. In order to find expressions for i_d and i_q, the flux linkages in Eqn. (8.3) are replaced by expressions derived from Eqns. (4.19) and (4.22). Hence Eqns. (8.3), when transformed into the Laplace domain, are

$$\begin{bmatrix} 0 \\ -\dfrac{U_m}{p} \end{bmatrix} = \begin{bmatrix} R_a + \dfrac{pX_d(p)}{\omega_0} & X_q(p) \\ -X_d(p) & R_a + \dfrac{pX_q(p)}{\omega_0} \end{bmatrix} \begin{bmatrix} \bar{i}_d \\ \bar{i}_q \end{bmatrix} \tag{8.4}$$

8.1.2 Solution of the short-circuit armature current

Eliminating $\bar{i}_q$ from Eqn. (8.4)

$$\frac{U_m}{p} = \left[p^2 + \omega_0^2 + p\omega_0 R_a \left(\frac{1}{X_d(p)} + \frac{1}{X_q(p)}\right) + \frac{\omega_0^2 R_a^2}{X_d(p)X_q(p)}\right] \frac{X_d(p)}{\omega_0^2}\,\bar{i}_d \tag{8.5}$$

The expression in the square brackets can be simplified by making use of the fact that R_a is small. The term in R_a^2 can be neglected entirely, and, in the term in R_a, $X_d(p)$ and $X_q(p)$ can be simplified by neglecting the resistances R_f, R_{kd}, R_{kq}. This is equivalent to replacing in this term all the factors of the form $(1 + Tp)$ by Tp. With this approximation, $X_d(p)$ and $X_q(p)$ reduce to the subtransient reactances X_d'' and X_q'' (see p. 89).

Then

$$\bar{i}_d = \frac{\omega_0^2}{p(p^2 + 2\alpha p + \omega_0^2)} \cdot \frac{U_m}{X_d(p)} \tag{8.6}$$

where

$$\alpha = \frac{\omega_0 R_a}{2}\left(\frac{1}{X_d''} + \frac{1}{X_q''} = \frac{1}{T_a}\right)$$

The partial fraction expansion of $1/X_d(p)$, given by Eqn. (4.36) is substituted into Eqn. (8.6):

$$\bar{i}_d = \frac{\omega_0^2}{(p^2 + 2\alpha p + \omega_0^2)}\left[\frac{1}{X_d} + \left(\frac{1}{X_d'} - \frac{1}{X_d}\right)\frac{T_d'p}{(1 + T_d'p)} + \left(\frac{1}{X_d''} - \frac{1}{X_d'}\right)\frac{T_d''p}{(1 + T_d''p)}\right]\frac{U_m}{p} \tag{8.7}$$

Similarly, neglecting R_a,

$$\bar{i}_q = \frac{-p\omega_0}{(p^2 + 2\alpha p + \omega_0^2)}\left[\frac{1}{X_q} + \left(\frac{1}{X_q''} - \frac{1}{X_q}\right)\frac{T_q''p}{(1 + T_q''p)}\right]\frac{U_m}{p} \tag{8.8}$$

The inverse Laplace transformation Eqns. (13.12) and (13.13) are applied to each term of Eqns. (8.7) and (8.8), putting a successively equal to 0, $1/T_d'$, $1/T_d''$, and $1/T_q''$.

Hence, approximately,

$$i_d = U_m\left[\frac{1}{X_d} + \left(\frac{1}{X_d'} - \frac{1}{X_d}\right)\epsilon^{-t/T_d'} + \left(\frac{1}{X_d''} - \frac{1}{X_d'}\right)\epsilon^{-t/T_d''}\right] - \frac{U_m}{X_d''}\cdot\epsilon^{-\alpha t}\cdot\cos\omega_0 t \tag{8.9}$$

$$i_q = \frac{-U_m}{X_q''}\epsilon^{-\alpha t}\cdot\sin\omega_0 t. \tag{8.10}$$

Substitution of the values of i_d and i_q in the transformation equations (4.5) with $\theta = (\omega_0 t + \lambda)$ gives the expressions for i_a, i_b and i_c:

$$\begin{aligned} i_a = U_m &\left[\frac{1}{X_d} + \left(\frac{1}{X_d'} - \frac{1}{X_d}\right)\epsilon^{-t/T_d'} + \left(\frac{1}{X_d''} - \frac{1}{X_d'}\right)\epsilon^{-t/T_d''}\right]\cos(\omega_0 t + \lambda) \\ &- \frac{U_m}{2}\left(\frac{1}{X_d''} + \frac{1}{X_q''}\right)\cdot\epsilon^{-t/T_a}\cdot\cos\lambda \\ &- \frac{U_m}{2}\left(\frac{1}{X_d''} - \frac{1}{X_q''}\right)\cdot\epsilon^{-t/T_a}\cdot\cos(2\omega_0 t + \lambda) \end{aligned} \tag{8.11}$$

The values of i_b and i_c are obtained by replacing λ by $(\lambda - 2\pi/3)$ and $(\lambda - 4\pi/3)$ respectively in the expression for i_a.

8.1.3 The short-circuit torque

In order to find an expression for the torque, it is necessary to determine the quantities ψ_α and ψ_q. Before the short-circuit the values are, from Eqns. (4.15):

$$\psi_{d0} = -\frac{U_m}{\omega_0},$$

$$\psi_{q0} = 0.$$

After the short circuit the values are:

$$\psi_d = \psi_{d0} + \psi_d',$$

$$\psi_q = \psi_{q0} + \psi_q',$$

where ψ_d' and ψ_q' are the superimposed quantities, for which Eqn. (8.3) holds. Using Eqns. (4.19) and (8.6),

$$\bar{\psi}_d' = \frac{X_d(p)}{\omega_0}\,\bar{i}_d = \frac{\omega_0}{(p^2 + 2\alpha p + \omega_0^2)}\,\frac{U_m}{p}$$

so that applying the inverse Laplace transformation Eqn. (13.12), with $a = 0$,

$$\psi_d' = \frac{U_m}{\omega_0}\,(1 - \epsilon^{-t/T_a} \cdot \cos\omega_0 t).$$

Hence

$$\psi_d = -\frac{U_m}{\omega_0} \cdot \epsilon^{-t/T_a} \cdot \cos\omega_0 t. \tag{8.12}$$

Similarly

$$\psi_q = -\frac{U_m}{\omega_0}\,\epsilon^{-t/T_a} \cdot \sin\omega_0 t. \tag{8.13}$$

The two flux waves represented by ψ_d and ψ_q combine to form a forward rotating flux wave, which rotates at speed ω_0 and is therefore stationary relative to the armature. Its magnitude decays with the armature time constant T_a, which depends on the armature resistance. Thus the effect of the short-circuit can be explained by imagining that the flux, which rotates relatively to the armature during normal operation, is frozen in position relative to the armature at the instant of short-circuit, and then decays with time constant T_a.

The torque is obtained by substituting the expressions for ψ_d, ψ_q, i_d, i_q in Eqn. (4.16) and rearranging the terms:

$$M_e = U^2 \cdot \epsilon^{-t/T_a} \cdot \left[\frac{1}{X_d} + \left(\frac{1}{X_d'} - \frac{1}{X_d}\right) \epsilon^{-t/T_d'} + \left(\frac{1}{X_d''} - \frac{1}{X_d'}\right) \epsilon^{-t/T_d''}\right] \sin \omega_0 t + \frac{U^2}{2} \cdot \epsilon^{-2t/T_a} \cdot \left(\frac{1}{X_q''} - \frac{1}{X_d''}\right) \sin 2\omega_0 t \qquad (8.14)$$

where U is the r.m.s. voltage.

The double frequency torque represented by the second term of Eqns. (8.14) is relatively small. Hence the principal component of torque oscillates at normal frequency and has an initial amplitude U^2/X_d''.

In practice there also exists a unidirectional component of torque due to winding resistive losses, but such a term does not appear in Eqn. (8.14) because of the simplifying assumptions made earlier on p. 162. For further details about the unidirectional torques the reader is referred to [24].

8.1.4 The field current after the short-circuit

The field current before the short-circuit is given by Eqns. (7.3), with $i_d = i_q = 0$.

$$i_{f0} = -\frac{u_{q0}}{X_{md}} = -\frac{U_m}{X_{md}}$$

The field current after the short-circuit is obtained by adding to i_{f0} the superimposed current i_f' determined by solving the equations already used to calculate the armature current. A relation between i_f' and i_d can be obtained by eliminating i_{kd}

from the second and third Eqns. (4.25) with $u_f = 0$. $\bar{i}_d$ is replaced by Eqn. (8.6).

$$\bar{i_f'} = \frac{L_{md}p(1 + T_{kd}p)}{R_f(1 + T_{d0}'p)(1 + T_{d0}''p)} \bar{i}_d$$

$$= \frac{(1 + T_{kd}p)}{(1 + T_d'p)(1 + T_d''p)} \cdot \frac{\omega_0 p}{(p^2 + 2\alpha p + \omega_0^2)} \cdot \frac{X_{md}}{X_d R_f} \cdot \frac{U_m}{p} \tag{8.15}$$

Now

$$\frac{(1 + T_{kd}p)}{(1 + T'p)(1 + T''p)} = \frac{1}{T_d'}\left[\frac{T_d'}{1 + T_d'p} - \left(1 - \frac{T_{kd}}{T''}\right)\frac{T_d''}{(1 + T_d''p)}\right]$$

approximately, if $T_d' \gg T_d''$, T_{kd}.

The solution is obtained by using the inverse Laplace transformation Eqn. (13.12), putting a successively equal to $1/T_d'$, $1/T_d''$.

$$i_f' = \frac{-U_m X_{md}}{\omega_0 T_d' R_f X_d}\left[\epsilon^{-t/T_d'} - \left(1 - \frac{T_{kd}}{T_d''}\right)\epsilon^{-t/T_d''} - \frac{T_{kd}}{T_d''}\epsilon^{-t/T_a} \cdot \cos \omega_0 t\right]$$

Now

$$-\frac{U_m X_{md}}{\omega_0 T_d' R_f X_d} = \frac{i_{f0}}{T_d'} \frac{X_{md}^2}{\omega_0 R_f X_d} = i_{f0}\left(\frac{T_{d0}' - T_d'}{T_d'}\right)$$

$$= i_{f0}\left(\frac{X_d - X_d'}{X_d'}\right).$$

Hence the total field current after the short-circuit is given by:

$$i_f = i_{f0} + i_f'$$

$$= i_{f0} + i_{f0}\left(\frac{X_d - X_d'}{X_d'}\right)\left[\epsilon^{-t/T_d'} - \left(1 - \frac{T_{kd}}{T_d''}\right)\epsilon^{-t/T_d''} - \frac{T_{kd}}{T_d''}\epsilon^{-t/T_a} \cdot \cos \omega_0 t\right]. \tag{8.16}$$

8.2 The analysis of short-circuit oscillograms

8.2.1 Oscillograms of armature currents

As already mentioned, a sudden short-circuit test, taken by short-circuiting simultaneously the three phases of a synchronous generator when running unloaded at rated voltage, is an accepted method of checking its transient characteristics. From the oscillograms of the armature currents the principal transient reactances and time constants can be calculated.

Fig. 8.1 shows the curves of armature current after a short-circuit on a 30 MVA synchronous condenser. The curves agree very closely with the expression for the current in phase A given in Eqn. (8.11) and the corresponding expressions for the currents in phases B and C.

In Fig. 8.1 envelope lines are drawn through the peaks of the

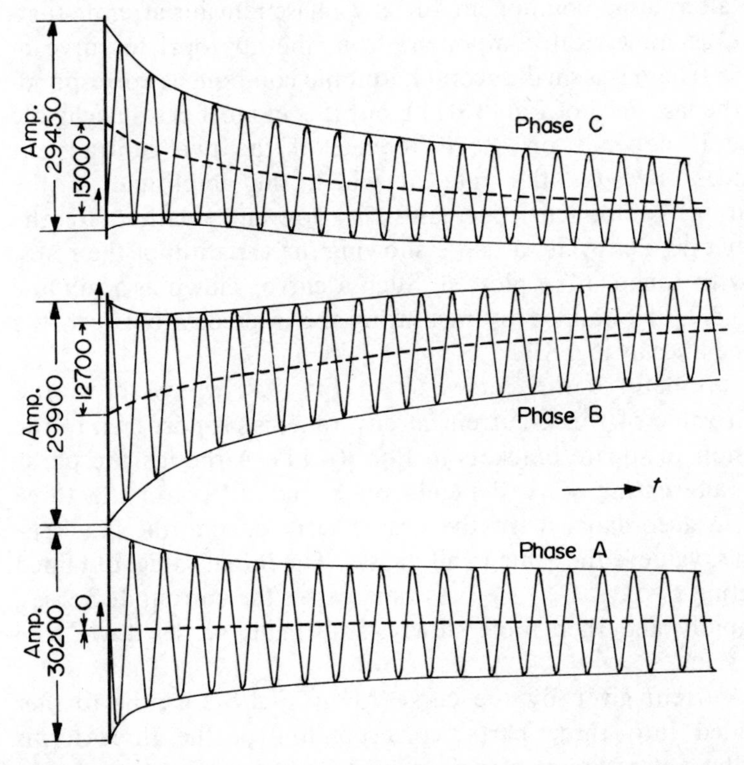

Fig. 8.1 Oscillograms of the armature currents after a short-circuit.

alternating current waves, and dotted lines are drawn half-way between the envelope curves. Thus the current can be divided into a unidirectional or *asymmetrical component*, and an *alternating component* of supply frequency. Both components start with a certain *initial value* and become less as time proceeds. The initial values of the alternating components are determined by producing the envelope curves back to zero time.

The dotted lines give the asymmetrical currents expressed by the fourth term of Eqn. (8.11) and by the fourth terms of the corresponding expressions for i_b and i_c. The initial values are different in the three phase A, B and C, being proportional to $\cos \lambda$, $\cos(\lambda - 2\pi/3)$ and $\cos(\lambda - 4\pi/3)$ respectively, but the three currents all decay to zero with the same time constant T_a. By replotting the dotted curves on semi-log paper (see [39]) the reactance X_m, the angle λ, and the time constant T_a, can be accurately determined.

The alternating component for any phase remains after deducting the asymmetrical component from the appropriate curve in Fig. 8.1. There is a small second harmonic component corresponding to the last term of Eqn. (8.11), but it is in most cases negligible because it depends on the difference of the two subtransient reactances. Because the rate at which the amplitude of the alternating component decreases is slow in relation to the fundamental a.c. cycle, a curve showing the variation of the r.m.s. value with time can be plotted. Such a curve, shown as a full line in Fig. 8.2, is obtained by measuring the intercepts between the envelope lines in Fig. 8.1.

The ordinate of the curve AB in Fig. 8.2, which shows the per-unit value of r.m.s. current at any time, is proportional to the expression in square brackets in Eqn. (8.11). Although the phase of the alternating wave depends on λ and differs in the three phases in accordance with the cosine term outside the brackets, the r.m.s. value is the same in all phases. The initial value, obtained by putting $t = 0$, is U_m/X_d''. As time passes the current decreases, first rapidly and then more slowly, and finally settles down to a steady value.

The current given by the curve AB in Fig. 8.2 can be further subdivided into three parts, corresponding to the three terms inside the square brackets in Eqn. (8.11). The steady short-circuit current, indicated by the dotted line EF, is U_m/X_d. The *transient*

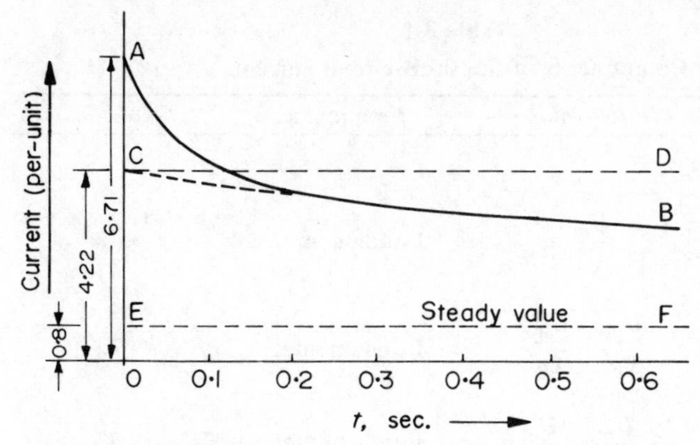

Fig. 8.2 Alternating component of the short-circuit armature current.

component, shown by the dotted line CB, has an initial value $EC = U_m/X_d' - U_m/X_d$, and decays with the time constant T_d'. The *subtransient component*, given by the intercept between AB and CB, has an initial value $CA = U_m/X_d'' - U_m/X_d'$ and decays with time constant T_d''. The five components of the short-circuit current are listed in Table 8.1.

Thus the values of the direct-axis transient and subtransient reactances and time constants can be deduced from the short-circuit oscillograms. The analysis of the short-circuit oscillogram brings out the physical meaning of the subtransient and transient reactances. The subtransient reactance X_d'' is the effective reactance of the generator determining the initial value of the alternating component of the short-circuit current. The transient reactance X_d' is the effective reactance determining the initial current which would flow if the rapidly decaying subtransient component were not present.

Figs. 8.1 and 8.2 are taken from [15] in which the results of a numerical check are given. The practical analysis is made by replotting the intercept between EF and AB on semi-logarithmic paper.

8.2.2 Oscillogram of field current

Fig. 8.3 shows the oscillogram of the field current taken at the same time as the armature current curves of Fig. 8.1. Envelope lines are drawn as before, and a dotted line is drawn half-way between the envelope lines. The total current is thus shown to

Table 8.1
Components of the short-circuit current

Component	*Initial value*	*Frequency*	*Time constant*
Alternating components			
Steady	$\frac{U_m}{X_d}$	Fundamental	
Transient	$\frac{U_m}{X_d'} - \frac{U_m}{X_d}$	Fundamental	T_d'
Subtransient	$\frac{U_m}{X_d''} - \frac{U_m}{X_d'}$	Fundamental	T_d''
Other components			
Asymmetrical	$\frac{U_m}{X_m} \cos \lambda$	Zero	T_a
Second harmonic	$\frac{U_m}{X_n}$	Double fundamental	T_a

consist of a undirectional component given by the dotted line, and an alternating component.

The alternating component corresponds to the last term of Eqn. (8.16). It is of fundamental frequency and decays with time constant T_a. The unidirectional component starts with the steady value i_{f0}, rises suddenly at the instant of short-circuit and follows the dotted curve until it finally returns to the steady value. This component corresponds to the first three terms of Eqn. (8.16).

The dotted curve of Fig. 8.3 derived from the oscillogram is replotted as the dotted curve in Fig. 8.4. Fig. 8.4 shows also the

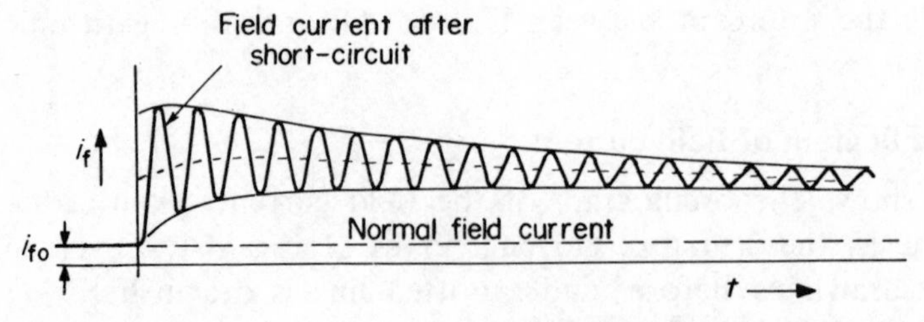

Fig. 8.3 Oscillogram of the field current after a short-circuit.

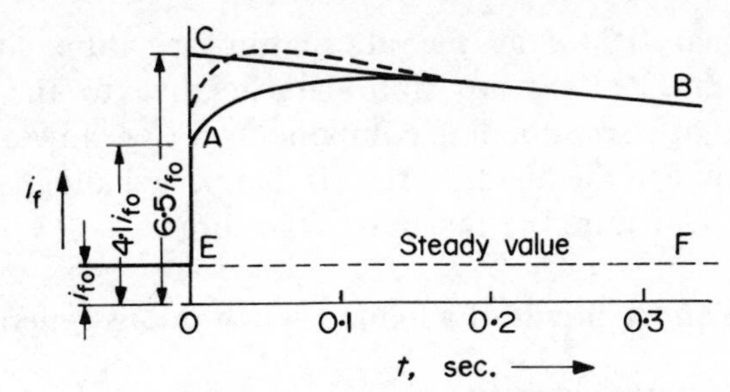

Fig. 8.4 Unidirectional component of the field current.

curve AB calculated from the first three terms of Eqn. (8.16) and the curve CB corresponding to the first two terms. The unidirectional field current given by the curve AB may be built up from three parts:

1. Steady field current i_{f0} given by the line EF.
2. Transient component given by the intercept between EF and CB. This part decays with time constant T_d'.
3. Subtransient component given by the intercept between CB and AB. This part is deducted from the curve CB and decays with time constant T_d''.

It can be seen that there is considerable discrepancy between the measured and calculated curves of field current. As explained in Section 4.7 the difference can be attributed to the assumption that the mutual inductance between field and damper is the same as that between field and armature. A detailed appraisal of the error introduced by this assumption is given in [42, 66 and 67].

The value of OC is $i_{f0}(X_d/X_d')$. This gives an indication of the rise in field current, but the peak current is actually greater because the oscillating component must be added to the curve CB.

The four components of the field current correspond quite closely to the four components of the direct-axis current i_d in Eqn. (8.9), both as regards the time constant of decay and whether the component is unidirectional or alternating. They also correspond to the components of the phase current i_a in Eqn. (8.11), but in this case a unidirectional component in the field current corresponds to an alternating component in the armature current and vice versa. In each case the corresponding components in the windings produce m.m.f.'s of the same type; for example, the unidirectional field and the alternating armature current both produce m.m.f.'s that are stationary relative to the field. The magnitudes of corresponding components differ, however, because of the current in the quadrature-axis damper winding, and because of the m.m.f. required to magnetize the core.

8.3 Short-circuit of a loaded synchronous generator

8.3.1 The armature current

For a synchronous generator which is short-circuited when it is operating on load, the same equations apply as for an unloaded machine, but the steady values of voltage and current before the short-circuit are different. The axis voltages u_{d0} and u_{q0} are given by Eqns. (7.3), and for a normal loaded condition u_{d0} is not zero. As on p. 161, the symbols U_{md} and U_{mq} are used to distinguish these steady voltages before the short-circuit.

The method of superposition can be used as before to determine the change of current resulting from the sudden application of voltages $-U_{md}$ and $-U_{mq}$. The appropriate equations are therefore:

$$\begin{bmatrix} \dfrac{-U_{md}}{p} \\ \dfrac{-U_{mq}}{p} \end{bmatrix} = \begin{bmatrix} R_a + \dfrac{pX_d(p)}{\omega_0} & X_q(p) \\ -X_d(p) & R_a + \dfrac{pX_q(p)}{\omega_0} \end{bmatrix} \begin{bmatrix} \bar{i}_d{}' \\ \bar{i}_q{}' \end{bmatrix} \tag{8.17}$$

The total short-circuit current i_a is obtained by adding the current determined from these equations to the original current i_{a0}, and is thus the sum of three quantities:

1. The original steady current i_{a0}.
2. The current i_{a1} obtained by solving Eqns. (8.17) with $U_{md} = 0$. The expression for i_{a1} is given by Eqn. (8.11).
3. The current i_{a2} is obtained by solving Eqns. (8.17) with $U_{mq} = 0$. The expression for i_{a2} similar to that in Eqn. (8.11), except that the roles of the direct and quadrature axes are interchanged. The components correspond to those of i_{a1},

except that there is no transient component because there is no field winding on the quadrature axis.

$$i_{a2} = -U_{md}\left[\frac{1}{X_q} + \left(\frac{1}{X_q''} - \frac{1}{X_q}\right)\epsilon^{-t/T_q''}\right] \sin(\omega_0 t + \lambda)$$

$$+ \frac{U_{md}}{2}\left(\frac{1}{X_d''} + \frac{1}{X_q''}\right)\epsilon^{-t/T_a} . \sin\lambda \qquad (8.18)$$

$$- \frac{U_{md}}{2}\left(\frac{1}{X_d''} - \frac{1}{X_q''}\right)\epsilon^{-t/T_a} . \sin(2\omega_0 t + \lambda)$$

Thus the theory explained in Section 8.1 can be used to obtain a complete solution for the current following a sudden three-phase short-circuit from any condition of steady operation. The field current i_f and the flux linkages ψ_d and ψ_q can also be determined in a similar way, using the principle of superposition.

8.3.2 The torque

The torque can be calculated by substituting the complete expressions for i_d, i_q, ψ_d, ψ_q, in Eqn. (4.16). It is found that the torque consists of oscillatory components and a unidirectional component but the latter is zero if the winding resistive losses are neglected [24].

The unidirectional component of torque always tends to decelerate the rotor while the oscillatory component alternately decelerates and accelerates it. As a result, there is an appreciable variation of speed, so that the results obtained by assuming constant speed are only approximate. They are accurate enough in practice for determining the performance immediately after the short-circuit, but for a longer term computation, particularly if the behaviour after clearing the fault is to be determined, the step-by-step method discussed in Chapter 9 must be used. Fig. 9.2, which was calculated by this method, shows clearly the unidirectional and oscillating components of torque, both before and after clearance.

8.4 Unsymmetrical short-circuit of a synchronous generator

When the circuits to which a generator is connected are not balanced between the three phases the conditions are much more complicated than when the circuits are balanced. It has already been shown in Section 7.3 that during steady operation negative-sequence applied voltages give rise not only to negative-sequence currents but also to third harmonic currents. Similarly negative-sequence currents give rise to third harmonic voltages. In the general case, when the voltages and currents are not specified, but are related to each other in a manner determined by the external circuits, both voltages and currents must be expressed by an infinite series of harmonics. Under transient conditions the voltage and current can be split up into components which decay at different rates, but each component contains an infinite series of harmonics.

With the assumption that the speed remains unchanged at the synchronous speed ω_0, Eqns. (4.15) give:

$$\left.\begin{aligned} u_d &= \dot{\psi}_d + \omega_0 \psi_q + R_a i_d \\ u_q &= -\omega_0 \psi_d + \dot{\psi}_q + R_a i_q \end{aligned}\right\} \tag{8.19}$$

The equations for the axis quantities are thus of a linear form, but the values of u_d and u_q are not known explicitly, as they are for a symmetrical short-circuit. There are additional relations between the axis voltages and currents depending on the external connections but they are in such a form that the ordinary operational methods cannot be used. A complete numerical solution can of course be obtained with a computer.

Complete analytical solutions for the principal unbalanced short-circuit conditions have, however, been obtained. For more general conditions, when the generator is connected to an external system, accurate analytical solutions are not possible but approximate methods similar to those explained in Section 8.5 can be used. For unbalanced conditions the method of symmetrical components is used and involves the assumption that all harmonics can be neglected.

The analysis of a single-phase short-circuit on a synchronous generator was first given by Doherty and Nickle [5], who derived expressions for the transient currents in the armature and field circuits and verified them by means of oscillographic tests. A full theoretical treatment of the three alternative types of short-circuit of a three-phase generator (line-to-line, line-to-neutral, and double-line-to-neutral) is given by Concordia [16]. For each case expressions are given for the transient torque and the open phase voltages as well as for the currents. The method used by these authors is to derive the initial values of the components of the currents by approximate methods and to estimate a time constant appropriate to each component.

A more rigorous and complete analysis of unbalanced conditions on synchronous machines is given in [19], which uses the Laplace transform method. This work provides a means of calculating the harmonics during steady unbalanced operation and gives a general formulation of the transient problem. The solutions obtained for the short-circuit conditions confirm the results of [5] and [16].

In order to indicate the nature of the problem, the particular case of a line-to-line short-circuit is considered in the next few pages. The discussion merely gives a statement of the equations and of the expression obtained for the armature current. For a more detailed treatment the references quoted above should be consulted.

8.4.1 Line-to-line short-circuit

The phase voltages and currents obey the following relations if phases B and C in series are short-circuited when the generator is running on open circuit.

$$i_a = 0,$$
$$i_b = -i_c,$$
$$u_b = u_c.$$

Using the transformation equations (4.5) and (4.7), the following relations between the axis quantities are obtained:

$$\left.\begin{aligned} i_d \cos\theta + i_q \sin\theta &= 0 \\ u_d \sin\theta_q - u_q \cos\theta &= 0 \end{aligned}\right\}, \quad (8.20)$$

where $\theta = \omega_0 t + \lambda$.

Eqns. (8.20), together with (8.19), (4.17), (4.20), and the known initial conditions, determine the voltages and currents. The solution is, however, difficult because the coefficients now include functions of time, and the principal advantage obtained by transforming to axis quantities is thereby lost. The solutions given in the references do not in fact use the axis quantities, but either determine the three phase currents and voltages directly or introduce equivalent two-phase quantities (α-β components). Whatever method is used for formulating the equations, the solution is much more difficult than for the symmetrical condition.

The expression for the short-circuit current obtained by solving the equations is:

$$i_{\mathrm{b}} = \sqrt{3}U\left[\frac{1}{X_{\mathrm{d}}+X_2} + \left(\frac{1}{X_{\mathrm{d}}'+X_2} - \frac{1}{X_{\mathrm{d}}+X_2}\right)\epsilon^{-t/T_{\mathrm{d}}'(l-l)}\right.$$
$$\left. + \left(\frac{1}{X_{\mathrm{d}}''+X_2} - \frac{1}{X_{\mathrm{d}}'+X_2}\right)\epsilon^{-t/T_{\mathrm{d}}''(l-l)}\right] \sum_{\mathrm{n}=0}^{\infty} (-\mathrm{b})^{\mathrm{n}} \cos(2n-1)\theta$$
$$+ \frac{\sqrt{3}U \sin\lambda}{X_2} \cdot \epsilon^{-t/T_{\mathrm{a}(l-l)}} \left[\frac{1}{2} + \sum_{\mathrm{n}=1}^{\infty} (-b)^n \cos 2n\theta\right] \tag{8.21}$$

where

$$\left.\begin{aligned}
X_2 &= \sqrt{X_{\mathrm{d}}''X_{\mathrm{q}}''}\\
b &= \frac{\sqrt{X_{\mathrm{q}}''} - \sqrt{X_{\mathrm{d}}''}}{\sqrt{X_{\mathrm{q}}''} + \sqrt{X_{\mathrm{d}}''}}\\
T_{\mathrm{d}}'(l-l) &= \frac{X_{\mathrm{d}}'+X_2}{X_{\mathrm{d}}+X_2}\,T_{\mathrm{d0}}'\\
T_{\mathrm{d}}''(l-l) &= \frac{X_{\mathrm{d}}''+X_2}{X_{\mathrm{d}}'+X_2}\,T_{\mathrm{d0}}''\\
T_{\mathrm{a}(l-l)} &= \frac{X_2}{\omega_0 R_{\mathrm{a}}}\,.
\end{aligned}\right\} \tag{8.22}$$

The quantities not defined here are given in Section 4.6.1.

The steady current after all transient components have died away is given by the following harmonic series:

$$[i_{\mathrm{b}}]_{(\mathrm{steady})} = \sqrt{3}U\left(\frac{1}{X_{\mathrm{d}}+X_2}\right) \sum_{n=0}^{\infty} (-b)^{\mathrm{n}} \cos(2n-1)\theta\,. \tag{8.23}$$

The form of the expression in Eqn. (8.21) for the current after an unbalanced short-circuit is similar to the corresponding expression in Eqn. (8.11) under balanced conditions, except that the first part contains an infinite series of odd harmonics and the second part contains an infinite series of even harmonics. The magnitudes of the harmonics depend on the quantity b, which is zero if the two subtransient reactances X_d'' and X_q'' are equal. For an approximate analysis it is permissible to neglect the harmonics as well as the asymmetrical component included in the second part of Eqn. (8.21). The fundamental alternating component is:

$$[i_b]_{(alt)} = \sqrt{3}U \left[\frac{1}{X_d + X_2} + \left(\frac{1}{X_d' + X_2} - \frac{1}{X_d + X_2} \right) \epsilon^{-t/T_d'(l-l)} + \left(\frac{1}{X_d'' + X_2} - \frac{1}{X_d' + X_2} \right) \epsilon^{-t/T_d''(l-l)} \right] \cos(\omega_0 t + \lambda). \tag{8.24}$$

8.4.2 Steady, transient and subtransient components of current

The current given by Eqn. (8.24) is of the same form as the alternating component, given by the first part of Eqn. (8.11) for the current after a symmetrical short-circuit. It consists of a steady component, a transient component and a subtransient component, but the values of the various reactances differ from those in the expression for the symmetrical short-circuit current.

The fundamental of the steady current in phase B during a sustained line-to-line short-circuit between phases B and C after all the transient components have died away is:

$$[i_b]_{st} = \frac{\sqrt{3}U}{X_d + X_2} \cos(\omega_0 t + \lambda). \tag{8.25}$$

It can readily be seen that this result agrees with that given by the theory of symmetrical components, if the positive-sequence reactance is X_d, and the negative-sequence reactance is X_2.

The initial value of the current given by Eqn. (8.24), with the subtransient component included, is:

$$[i_b]'' = \frac{\sqrt{3}U}{X_d'' + X_2} \cos(\omega_0 t + \lambda). \tag{8.26}$$

The initial value, if the subtransient component is neglected, is

$$[i_\mathrm{d}]' = \frac{\sqrt{3}U}{X_\mathrm{d}' + X_2} \cos(\omega_0 t + \lambda). \tag{8.27}$$

These currents agree with those given by the symmetrical component method, if the positive-sequence reactances are taken to be X_d'' and X_d' respectively and the negative-sequence reactance to be X_2.

It thus appears that, for a line-to-line short-circuit, the steady and the initial transient values of the fundamental alternating component of the current can be calculated in a simple manner by the method of symmetrical components. The appropriate positive-sequence reactances are the same as those for the symmetrical short-circuit, and the negative-sequence reactance X_2 is that given in Eqns. (8.22).

The time constants $T_\mathrm{d}'{}_{(l-l)}$ and $T_\mathrm{d}''{}_{(l-l)}$ of the transient and subtransient components differ from the corresponding quantities T_d' and T_d'' for the symmetrical short-circuit current. The modified time constants have the same values as if there was a series reactance X_2 (see p. 90).

8.4.3 General treatment of unbalanced fault conditions

If a similar analysis is made for the line-to-neutral and double-line-to-neutral short-circuit conditions, it is found that the fundamental currents can also be determined by the symmetrical component method. The appropriate positive-sequence reactances are X_d, X_d', X_d'', as before, but the required values of negative-sequence reactance are different from that for the line-to-line short-circuit. The zero-sequence reactance X_z must also be used.

For more complicated unbalanced fault conditions of the kind encountered in power systems where several generators are connected through transformers and transmission lines, it is thus a reasonable assumption that the fundamental fault currents can be calculated by the method of symmetrical components. The principal doubt arises in the choice of the appropriate negative-sequence reactance X_2. The value of X_2 always lies between the two extreme values given by Eqns. (7.16) and (7.18) [14].

Compared with the symmetrical condition, for which the usual approximate methods of analysis are discussed in Section 8.5, the principal additional source of error is due to the presence of harmonic components. The magnitudes of the harmonics depend on the quantity b, which in turn depends on the ratio of the direct and quadrature-axis subtransient reactances. For the line-to-line short-circuit b has the value in Eqn. (8.22) and for other conditions the expression for b is different, but is of a similar type. When the two subtransient reactances X_d'' and X_q'' are equal, b is zero, and there are no harmonics. Moreover if $X_d'' = X_q''$, the two extreme values of X_2 are the same, as noted on p. 145.

If a synchronous generator is liable to operate with unbalanced currents, it is very desirable, in order to reduce the harmonics, that the two subtransient reactances shall be as nearly equal as possible. The results obtained by the method of symmetrical components are then sufficiently accurate for most practical purposes of power system analysis. When X_d'' and X_q'' are nearly equal, it is usually satisfactory to take the negative-sequence reactance X_2 to be equal to their arithmetic mean, as in Eqn. (7.18). If X_d'' and X_q'' differ considerably, the calculations become inaccurate, and there are liable to be other incidental troubles; for example, excessive voltage peaks may appear on an open phase.

8.5 System fault calculations

In Sections 8.1 to 8.3, complete solutions were derived for the current following a symmetrical short-circuit on a generator, assuming that the speed of the generator remained constant after the short-circuit and that certain resistances were negligible. Similar but more complicated problems arise when a short-circuit or other sudden change occurs at any point in a power system, which may include several generators connected together with transformers, transmission lines, and other apparatus. For such a system a complete solution can only be obtained by a step-by-step computation. A practical method can however be developed by making further simplifying assumptions.

The transient currents which flow after a sudden change in a power system can be divided into transient, subtransient, and asymmetrical components in the same way as the short-circuit currents considered earlier in this chapter. For many purposes it is not necessary to obtain a complete solution; for example, it may be permissible to neglect the asymmetrical component and determine only the alternating component. Simplified results of this kind are obtained if certain modifications are made to the basic assumptions, as explained below. Different assumptions are made according to which of the components are important in the particular problem.

For power-system analysis it is desirable for the sake of computer time (see Section 5.6) to reduce the order of the differential equations to the minimum needed to represent the bare essentials of the generator. This section shows how the simplifications, already discussed in Section 5.3, may be carried further by using only a constant driving voltage and a constant series impedance to represent the machine electrically.

8.5.1 Equivalent circuits

Equivalent circuits for the alternating components of current

The use of equivalent circuits for determining the transient performance of a generator after a sudden change depends on the assumption that the rate of change of the amplitudes of the alternating quantities is slow in relation to the a.c. cycle. This is equivalent to the assumption that the axis quantities change slowly; in particular, that the voltages $\dot{\psi}_d$ and $\dot{\psi}_q$, induced in either axis coil by the rate of change of the flux on that axis, can be ignored. With this assumption the solution of the short-circuit problem discussed in Sections 8.1 and 8.3, is considerably simplified. It is found that, if the above assumption is made, and if also the armature resistance is neglected, a solution of the equations gives correctly the alternating component of the current, but omits altogether the asymmetrical and second harmonic components [15].

If the direct- and quadrature-axis voltages are suddenly reduced by U_{md} and U_{mq} respectively, as assumed in Section 8.3, the armature voltage equations are Eqns. (8.17). If then $\dot{\psi}_d$ and $\dot{\psi}_q$ are put equal to zero, Eqns. (8.17) become, with $R_a = 0$:

$$\begin{bmatrix} -\dfrac{U_{md}}{p} \\ -\dfrac{U_{mq}}{p} \end{bmatrix} = \begin{bmatrix} \omega_0 \bar{\psi}_q \\ -\omega_0 \bar{\psi}_d \end{bmatrix} = \begin{bmatrix} & X_q(p) \\ -X_d(p) & \end{bmatrix} \begin{bmatrix} \bar{i}_d{}' \\ \bar{i}_q{}' \end{bmatrix} \qquad (8.30)$$

From Eqns. (4.36) and (8.30) the solution of i_d' as a function of time is:

$$i_d' = U_{mq}\left[\frac{1}{X_d} + \left(\frac{1}{X_d'} - \frac{1}{X_d}\right)\epsilon^{-t/T_d'} + \left(\frac{1}{X_d''} - \frac{1}{X_d'}\right)\epsilon^{-t/T_d''}\right] \quad (8.31)$$

and the corresponding component of the phase current i_{a1} using the transformation Eqn. (4.5) with $\theta = (\omega_0 t + \lambda)$, is given by

$$i_{a1} = U_{mq}\left[\frac{1}{X_d} + \left(\frac{1}{X_d'} - \frac{1}{X_d}\right)\epsilon^{-t/T_d'} + \left(\frac{1}{X_d''} - \frac{1}{X_d'}\right)\epsilon^{-t/T_d''}\right]\cos(\omega_0 t + \lambda) \quad (8.32)$$

Hence the solution obtained by assuming that the changes are slow gives only the unidirectional component of direct-axis current or the corresponding alternating component of armature phase current. For problems where the alternating component is the only one that matters, the work of obtaining the solution is significantly reduced.

The equivalent circuit (Fig. 4.5a), with $u_f = 0$, and with the value of ψ_d given by Eqn. (8.30), now becomes that shown in Fig. 8.5 (circuit Da). The current that flows in this circuit, when the voltage U_{mq}/ω_0 is applied, is the undirectional part of the direct-axis current i_d', and corresponds to the alternating component i_{a1} of the phase current.

In a similar way the alternating current component i_{a2}, due to the voltage U_{md} can be found. The operational solution from Eqn. (8.30) is:

$$\bar{i}_q' = -\frac{U_{md}}{pX_q(p)} = -\frac{(1 + T_{q0}''p)\,U_{md}}{(1 + T_q''p)\,X_q p} \quad (8.33)$$

The solution for i_q' as a function of time is:

$$i_q' = -U_{md}\left[\frac{1}{X_q} + \left(\frac{1}{X_q''} - \frac{1}{X_q}\right)\epsilon^{-t/T_q''}\right] \quad (8.34)$$

and the corresponding phase current i_{a2} is given by:

$$i_{a2} = -U_{md}\left[\frac{1}{X_q} + \left(\frac{1}{X_q''} - \frac{1}{X_q}\right)\epsilon^{-t/T_q''}\right]\sin(\omega_0 t + \lambda) \quad (8.35)$$

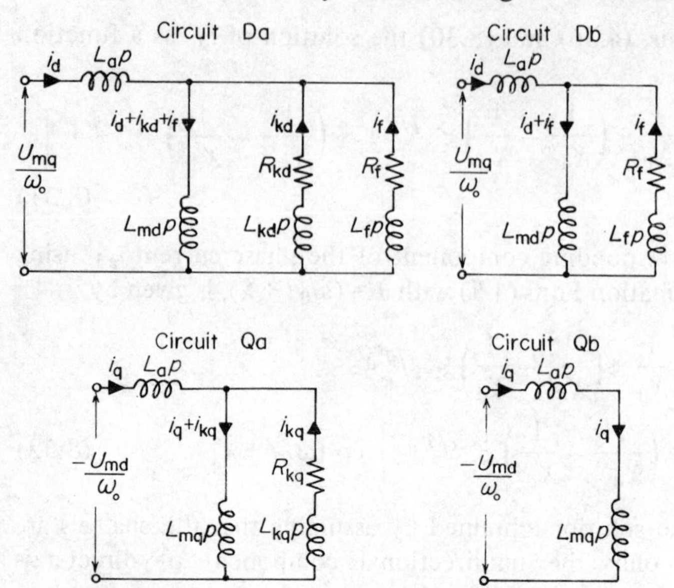

Fig. 8.5 Equivalent circuits for the alternating components of the current.
(a) Including the subtransient component.
(b) Neglecting the subtransient component.

The quadrature-axis equivalent circuit (Fig. 4.5b) now becomes that shown in Fig. 8.5 (circuit Qa), which is similar to the direct-axis circuit Da, except that there is no field winding branch.

Thus by means of the equivalent circuits Da and Qa of Fig. 8.5, the alternating component of the current which flows, after the voltage is suddenly changed, is obtained as the sum of three parts:

1. The original current i_{a0}.
2. The current i_{a1}, obtained from the current i_d, which flows in the direct-axis circuit Da when U_{mq}/ω_0 is applied.
3. The current i_{a2}, obtained from the current i_q, which flows in the quadrature-axis circuit Qa when $-U_{md}/\omega_0$ is applied.

The instantaneous values of the components of the applied voltage in phase A are given by Eqns. (8.2). Thus each of the currents given by Eqns. (8.32) and (8.35) lags 90° behind the voltage producing it, i_{a1} being produced by $(u_{a0})_q$, and i_{a2} by $(u_{a0})_d$.

Equivalent circuits for the alternating components of current, neglecting the subtransient part

For many purposes the subtransient components, which die away quickly, can be neglected when calculating the current following a short-circuit or other sudden change of voltage. The equivalent circuits can then be simplified to circuits Db for the direct axis, and Qb for the quadrature axis. With these simplifications the direct-axis current consists only of a steady component and a transient component, while the quadrature-axis current attains the final steady value immediately.

Equivalent circuits for the initial values of the alternating components

When the problem is such that it is only necessary to determine the initial value of the transient current, further simplifications can be made. The current decays because of the dissipation of energy in the resistance of the field and damper windings. If these resistances were zero, the initial current would be maintained indefinitely.

Considering first the direct-axis currents, the equations simplify to the following ordinary algebraic equations (containing no derivative operators), if the resistances of the field and damper windings are assumed to be zero.

$$\left.\begin{aligned} U_{mq} &= X_{md}i_f + X_{md}i_{kd} + (X_{md} + X_a)i_d \\ u_f &= 0 = (X_{md} + X_f)i_f + X_{md}i_{kd} + X_{md}i_d \\ u_{kd} &= 0 = X_{md}i_f + (X_{md} + X_{kd})i_{kd} + X_{md}i_d \end{aligned}\right\} \quad (8.36)$$

Eqns. (8.36) determine a constant value of i_d equal to the initial value U_{mq}/X_d'' of the expression in Eqn. (8.32). The corresponding phase current i_{a1} is an alternating current, which does not die away, given by:

$$i_{a1} = \frac{U_{mq}}{X_d''}\cos(\omega_0 t + \lambda).$$

The equivalent circuits of Fig. 8.6 can be used to determine the initial transient values. The current i_{a1} and the corresponding voltage u_{a2} are the instantaneous values of current and voltage represented by phasors I_d and U_q, such that I_d lags 90° behind

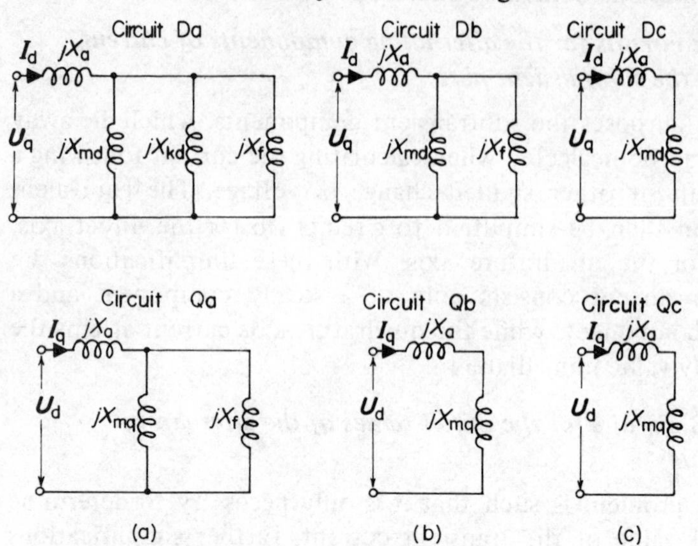

Fig. 8.6 Equivalent circuits for the initial and steady-state values of the current.
(a) Including the subtransient component.
(b) Neglecting the subtransient component.
(c) Steady-state condition.

U_q. Because of the relations between the axis quantities and the components of the phasors representing the phase values during steady operation (see p. 134), Eqns. (8.36) show that I_d is the current that flows in the simplified a.c. equivalent circuit Da of Fig. 8.6 when the applied voltage is U_q. The expression for X_d'' on p. 89 shows that the overall reactance of the circuit Da is the direct-axis subtransient reactance X_d''. Similarly the corresponding quadrature-axis circuit is Qa of Fig. 8.6, of which the overall reactance is the quadrature-axis subtransient reactance X_q''.

The currents given by the circuits Da and Qa of Fig. 8.6 are the initial currents, allowing for the effect of the damper winding. If the subtransient component is neglected, as it can be for some purposes, the circuits reduce to Db and Qb, in which the damper coils are omitted. The expression for X_d' on p. 89 shows that the overall reactance of circuit Db is the direct-axis transient reactance X_d', while that of circuit Qb is the quadrature-axis synchronous reactance X_q, which applies for the transient condition as well as for the steady state. The current obtained with these circuits is the initial value that would be obtained if the rapidly decaying subtransient component were not present. For example, referring to Fig. 8.2, the short-circuit current calculated from the equivalent circuit Db of Fig. 8.6 has a constant alternating value, for which the magnitude is given by the line CD, instead of the varying value given by the curve AB.

For the sake of completeness the steady-state equivalent circuits Dc and Qc are also given in Fig. 8.6. The overall reactances of these circuits are the synchronous reactances.

For the simplified conditions to which the use of the equivalent circuits apply, the alternating voltages and currents are assumed to have definite values at any instant, because, apart from the sudden initial change, the rate of change of the quantities is slow compared with the a.c. cycle. The sudden initial change is a quick change of the alternating value. When the additional assumption is made that all resistances are zero, the effect of a sudden change of the a.c. voltage at the alternator terminals is simply to change the current from one a.c. value to another. The change can be calculated from two components, one flowing in the direct-axis equivalent circuit when the quadrature-axis voltage is applied, and the other flowing in the quadrature-axis circuit when the direct-axis voltage is applied. If the resistances were really zero the current would remain at this value indefinitely, but in a practical case where the resistances are not zero, the current calculated by this method is the initial value. As time passes the current changes and finally attains the steady value.

The particular equivalent circuit to be used depends on the requirements of the problem. If the initial value of the current, including the subtransient component, is required, the circuits Da and Qa would be used. If, however, the subtransient component can be neglected the appropriate circuits are Db and Qb. Circuits Dc and Qc apply for steady conditions.

8.5.2 The constant-flux-linkage theorem

The constant-flux-linkage theorem, due to Doherty [1] has frequently been used for the study of transient phenomena in synchronous machines. The theorem depends on the well-known fact that the flux linkage with an inductive circuit having zero resistance cannot change, whatever may occur in other mutually coupled circuits. In the practical case, where resistance is present, it is still true that the flux linkage with an inductive circuit cannot change suddenly. If a sudden change, such as a short-circuit, occurs in a coupled circuit, the flux linkage with the closed circuit remains unaltered for the first instant, but changes in time to a final steady value as energy is absorbed in the resistance.

In applying the constant-flux-linkage theorem to the synchronous machine, the additional assumption is usually made that the alternating quantities change slowly so that the $\dot{\psi}_d$ and $\dot{\psi}_q$ terms in Eqn. (4.15) may be neglected. Without this additional assumption, the theorem leads to a solution containing an asymmetrical component as obtained by Doherty and Nickle in [5]. Furthermore, it is assumed that the volt drop across the armature resistance may be neglected in Eqn. (4.15). Under these conditions, Eqns. (4.15), (4.25) and (4.26) are combined [49] to yield:

$$\left.\begin{aligned} u_d &= X_q''i_q + u_d'' \\ u_q &= X_d''i_d + u_q'' \end{aligned}\right\} \quad (8.39)$$

where

$$\left.\begin{aligned} u_d'' &= \omega_0(X_q'' - X_a)(\psi_{kq}/X_{kq}) \\ u_q'' &= -\omega_0(X_d'' - X_a)\left(\frac{\psi_f}{X_f} + \frac{\psi_{kd}}{X_{kd}}\right) \end{aligned}\right\} \quad (8.40)$$

The two components u_d'' and u_q'' represent internal voltages which, because of the constant flux linkages ψ_f, ψ_{kd} and ψ_{kq}, remain unaltered for the first instant.

The constant-flux-linkage method therefore gives the initial value of the alternating component of current, exactly as obtained from the a.c. equivalent circuits of Fig. 8.6. For steady a.c. conditions the relationships of Section 7.1 are used to rearrange Eqn. (8.39) so that

$$\left.\begin{aligned} U_d - jX_q''I_q &= U_d'' \\ U_q - jX_d''I_d &= U_q'' \end{aligned}\right\} \quad (8.41)$$

The equivalent circuits show more clearly how the results are obtained and they are therefore used in the following sections in preference to the constant-flux-linkage method.

8.5.3 The two-axis phasor diagram for transient changes

The equivalent circuits Da and Qa of Fig. 8.6, of which the overall reactances are X_d'' and X_q'' respectively, give the relations between the change of voltage and the change of current in a generator after any sudden change. When the generator is connected to a power system the sudden change may be due to a fault or a switching operation at any point in the system. From the relations given by the equivalent circuits it is possible to construct a phasor diagram which can be used to determine the initial voltage and current after the change.

The results of Eqn. (8.41) could also be deduced from the equivalent circuits of Fig. 8.6 as follows. Let $\Delta \boldsymbol{U}_d$ and $\Delta \boldsymbol{U}_q$ be the sudden changes in the components of voltage, and $\Delta \boldsymbol{I}_d$ and $\Delta \boldsymbol{I}_q$ the corresponding changes in the components of current, assuming that the three phases remain balanced throughout. Then from circuits Da and Qa of Fig. 8.6,

$$\Delta \boldsymbol{U}_q = jX_d''\Delta \boldsymbol{I}_d$$
$$\Delta \boldsymbol{U}_d = jX_q''\Delta \boldsymbol{I}_q$$

Hence

$$\boldsymbol{U}_q - jX_d''\boldsymbol{I}_d = \text{constant} = C_d$$
$$\boldsymbol{U}_d - jX_q''\boldsymbol{I}_q = \text{constant} = C_q$$

where the constants C_d and C_q are $\boldsymbol{U}_d''$ and $\boldsymbol{U}_q''$ respectively as defined by Eqn. (8.41).

Thus for a short period after a sudden change (during which $\boldsymbol{U}$ and $\boldsymbol{I}$ change) brought about by a fault or the operation of a switch anywhere in the system, the voltages $\boldsymbol{U}_d''$ and $\boldsymbol{U}_q''$, obtained by adding the 'subtransient reactance drops' $(-jX_d''\boldsymbol{I}_d)$ and $(-jX_q''\boldsymbol{I}_q)$ to the components $\boldsymbol{U}_q$ and $\boldsymbol{U}_d$ of the terminal voltage (see Eqn. (8.41)) remain unaltered. The relations are shown on Fig. 8.7, which is a phasor diagram similar to Fig. 3.10, except that the armature resistance R_a is neglected. PL'' is the direct-axis subtransient reactance drop $(-jX_d''\boldsymbol{I}_d)$, and $L''N''$ is the quadrature-axis subtransient reactance drop $(-jX_q''\boldsymbol{I}_q)$. ON'' is $\boldsymbol{U}''$, the resultant of $\boldsymbol{U}_d''$ and $\boldsymbol{U}_q''$, and is called the *voltage behind subtransient reactance.* The voltage $\boldsymbol{U}''$ remains initially unaltered after any sudden change.

If the machine has no damper winding, or if the conditions of

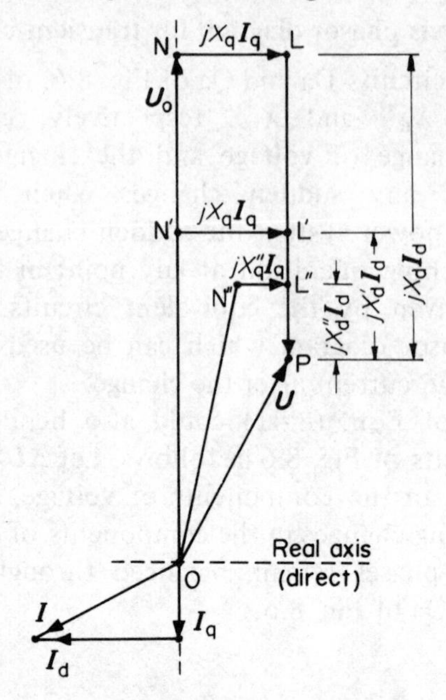

Fig. 8.7 Two-axis phasor diagram for the initial current and voltage.

the problem are such that the rapidly decaying subtransient components of voltage and current can be neglected, the relations between the sudden changes of voltage and current are given by the equivalent circuits Db and Qb of Fig. 8.6, of which the overall reactances are X_d' and X_q respectively.

In a similar way to the derivation of Eqn. (8.41) it can be shown that, without a damper winding,

$$\left.\begin{aligned} U_d - jX_q I_q &= U_d' = \text{constant} \\ U_q - jX_d' I_d &= U_q' = \text{constant} \end{aligned}\right\} \qquad (8.42)$$

Thus for the effective initial condition ignoring the subtransient components, the voltages U_q' and U_d', obtained by adding the 'transient reactance drops' $(-jX_d' I_d)$ and $(-jX_q I_q)$ to the components U_q and U_d of the terminal voltage, remain unaltered. In Fig. 8.7, PL' is the direct-axis transient reactance drop $(-jX_d' I_d)$ and $L'N'$ is the quadrature-axis synchronous reactance drop $(-jX_q I_q)$. ON' is U', the resultant of U_d' and U_q', and is called the *voltage behind transient reactance.* The voltage U' remains initially unaltered after any sudden change.

Fig. 8.7 also shows the steady-state phasor diagram, in which the synchronous reactance drops $(-jX_d I_d)$ and $(-jX_q I_q)$ are added to the terminal voltage. ON is U_0, the *voltage behind synchronous reactance*, which remains unaltered for any condition of steady operation.

The phasor diagrams for the two alternative types of transient condition are similar to the steady-state diagram, differing only in the values of reactance used. Hence the use of the phasor diagram to determine the initial currents after a sudden change follows the same lines as the steady-state two-axis method.

Simplified phasor diagrams for transient changes

When the generator is connected to a power system the calculations required by the two-axis theory become rather complicated. The method is greatly simplified if the assumption is made that each quadrature-axis reactance is equal to the corresponding direct-axis reactance, that is:

$X_q = X_d$ (uniform air-gap machine),
$X_q = X_d'$ (zero 'transient saliency'),
$X_q'' = X_d''$ (zero 'subtransient saliency').

Fig. 8.8 is a phasor diagram showing the result of these assumptions. The points N, N′ and N″ on the two-axis diagram (shown dotted) move to M, M′ and M″, which all lie on a straight line through P perpendicular to the current phasor *I*. *PM* is the synchronous reactance drop $(-jX_d I)$, *PM′* is the transient reactance drop $(-jX_d' I)$, and *PM″* is the subtransient reactance drop $(-jX_d'' I)$.

The assumption that $X_q = X_d$ for steady operation means that the generator is treated as a uniform air-gap machine. U_s, the voltage behind synchronous reactance, is then equal to OM, and the generator can be represented by a source of constant voltage U_s with the synchronous reactance X_d in series. U_s differs from U_0 because of the changed value of X_q.

For calculations relating to 'slow' transient changes, for which the subtransient components can be neglected, the assumption is that $X_q = X_d'$. The machine is said to have 'zero transient

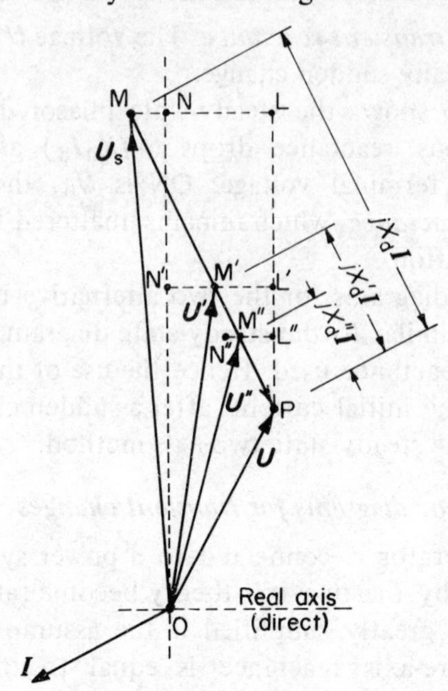

Fig. 8.8 Simplified phasor diagram for the initial current and voltage.

saliency', a condition which is quite different from that of having a uniform air-gap because of the effect of the field winding. The assumption is in fact more nearly true of a salient-pole generator than of a turbo-generator. With this assumption U', the voltage behind transient reactance, is equal to OM', and the generator can be represented by a source of constant voltage U' with the transient reactance $X_d{}'$ in series.

For the calculation of 'rapid' transients, where the initial values of voltage and current immediately after a sudden change are required, the approximate method assumes that the machine has 'zero subtransient saliency', or that $X_q{}'' = X_d{}''$. For most modern machines the assumption is a reasonable one, particularly as the subtransient reactance is small compared with the reactances in the external power system. U'', the voltage behind subtransient reactance, is equal to OM'' in Fig. 8.8, and the generator can be represented by a source of constant voltage U'' with the subtransient reactance $X_d{}''$ in series.

8.5.4 Application of the approximate methods to power-system transient analysis

The approximate methods discussed above provide the basis for simple means of determining the currents in a power system under certain conditions, since the generator can be replaced by its driving voltage and reactance appropriate to the particular condition. The system then becomes an ordinary network of lumped elements and any of the well-known methods of network analysis can be used. For extensive systems a computer is indispensable.

Fault and short-circuit conditions

After the occurrence of a fault or a short-circuit at any point of a power-system, the current immediately rises to a high value, but decays to a new steady value in a few seconds, as shown in Fig. 8.1. The rupturing capacity of a switch installed at the point must be sufficient to enable the circuit to be opened safely after any possible fault condition.

The transient current consists of an alternating component, whose magnitude does not depend on the instant of switching, and an asymmetrical component which displaces the wave for a short time after the fault. The initial value of the alternating component can be computed if each generator is represented by its subtransient reactance and a constant 'voltage behind subtransient reactance', which is taken to be the driving voltage. The voltage behind subtransient reactance is the voltage $\boldsymbol{U}''$ introduced on p. 187. The original steady current must be determined by a load flow computation in order to calculate the value of $\boldsymbol{U}''$ for each generator. The network diagram is then modified to represent the condition after the fault, assuming that the same value of $\boldsymbol{U}''$ is maintained. The r.m.s. current that flows at any point in the system under this condition is the initial value of the alternating component immediately after the fault.

The maximum peak current is greater than the peak of the alternating component because of the displacement due to the asymmetrical component, and, in the worst condition with the most favourable instant of switching, the value may be doubled. Because of the rapid decay of the wave, the theoretical initial value is never attained, as can be seen in Fig. 8.1 (phase C), and it is usually assumed, as specified in British Standard Specification No. 116, that the maximum peak is 1.8 times the peak value of the alternating current calculated in the manner described. Thus if I'' is the computed r.m.s. current in the network, the maximum peak current is $1.8\sqrt{2}\,I''$.

Transient stability

As explained in Section 9.1, the transient stability of a system after a large disturbance is determined by computing a swing curve for each generator. However, a rough determination of the transient stability of a machine, with constant field voltage and connected to an infinite bus, can be made by using the *equal area criterion* [11], which is based on the constant-flux-linkage theorem.

By this theorem (see Section 8.5.2) the voltage U' behind the transient reactances $X_d{}'$ on the direct axis and X_q on the quadrature-axis remains approximately constant for a period after a sudden change. Thus a transient power-angle curve can be drawn, similar to Fig. 3.11, but with U_0 replaced by U' and X_d replaced by $X_d{}'$ in Eqn. (3.8). The effective quadrature-axis reactance is X_q, because there is no field winding. Fig. 8.9 shows such a curve, in which the peak value is higher than in Fig. 3.11 and occurs at $\delta_g > 90°$, because $X_d{}' < X_q$. Fig. 8.9 is drawn for generator operation with power P_g and angle δ_g.

If, to take a simple example, the input power of a generator operating steadily at power P_1 and angle δ_1, is suddenly increased to P_2, the electrical power cannot change instantly and the surplus energy is stored in the kinetic energy of the rotating shaft mass. This causes an increase in speed, a rise in rotor angle and therefore a rise in electrical power. Fig. 8.9 shows how the electrical power varies with the rotor angle. At angle δ_2 there is a temporary equilibrium condition where the input shaft power balances the electrical power, but the speed is above synchronism and δ_g therefore rises further until, at angle δ_3, for which area A_1 = area A_2, the speed is back to synchronous speed. δ_3 is the highest value attained and after some oscillation, the machine settles down to the steady angle δ_2 corresponding to the power P_2. The equal area criterion states that the machine remains in synchronism if δ_3 does not exceed δ_4, the value on the downward sloping part of the

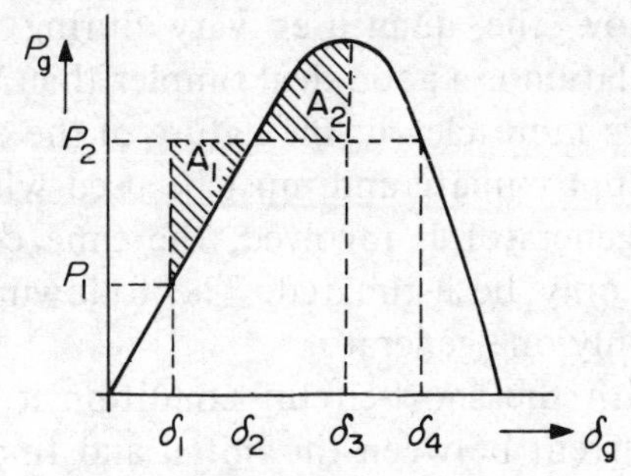

Fig. 8.9 Transient power-angle characteristic of a synchronous generator with transient saliency.

curve, corresponding to input power P_2. If δ_g exceeds δ_4 the machine continues to accelerate, because the electrical power is less than P_2, so that synchronism is lost and the system is unstable.

It is interesting to note that the assumptions made above are similar to those made in relation to Eqn. (5.25), but with the additional assumption that the input variables z_1 and z_2 are both constant.

The equal area criterion can be applied to many types of disturbance, of which the one discussed above is a simple example. It is featured in many textbooks and is useful, in spite of its poor accuracy, because it can help a student to understand the concept of stability. In the past, somewhat better accuracy was obtained by making a numerical computation, using a network analyser, based on the simplified generator representation, but it was difficult to allow for regulating devices. However, when a digital computer is available, there is no need to accept an inaccurate solution. Practical computations of transient stability should use an adequate representation of the generators in a system, together with their regulators and governors, as explained in Chapter 9.

8.6 Sudden load changes

The methods described in Section 8.5 provide means of determining the initial values of voltage and current in a system immediately after a sudden change, as well as the final values after a steady condition has been reached. It is often necessary to know approximately how the quantities vary during the intervening period. The calculation is a good deal simpler than the step-by-step process and shows more clearly the nature of the changes, but the method is only approximate and must be used with discretion. If more than one generator is involved, the time constants of the components can only be estimated. The following examples are concerned with only one generator.

By analogy with the short-circuit condition it is assumed that the change of current between the initial and final values can be divided into two component currents, each of which decays exponentially with a definite time constant. The problem is therefore to determine the magnitudes and time constants for each of these components.

8.6.1 Sudden application of load to a generator

If a load is suddenly applied to a generator running on open circuit with a constant excitation, the voltage falls and eventually reaches a new steady value. The starting of a large induction motor, suddenly switched on to the generator, is a common example of this condition. There is first a rapid change and then a more gradual one, as indicated by the full line on Fig. 8.10. If the load is a pure reactance, the solution is obtained by adding the external

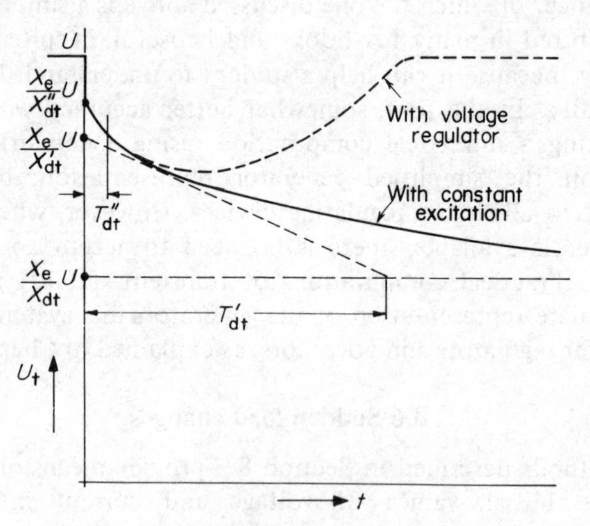

Fig. 8.10 Variation of the voltage after sudden application of a load.

reactance X_e to that of the generator and using the result already worked out in Section 8.1 for the short-circuit. The alternating component of the current that flows after the switch is closed, is from Eqn. (8.11):

$$i_{at} = \sqrt{2}U\left[\frac{1}{X_{dt}} + \left(\frac{1}{X_{dt}'} - \frac{1}{X_{dt}}\right)\epsilon^{-t/T_{dt}'} + \left(\frac{1}{X_{dt}''} - \frac{1}{X_{dt}'}\right)\epsilon^{-t/T_{dt}''}\right]\cos(\omega_0 t + \lambda), \tag{8.43}$$

where U is the r.m.s. voltage before the load is connected, and the additional suffix t indicates that $(X_a + X_e)$ must be used instead of X_a as explained on p. 90.

The r.m.s. voltage at the generator terminals is given by:

$$U_t = UX_e\left[\frac{1}{X_{dt}} + \left(\frac{1}{X_{dt}'} - \frac{1}{X_{dt}}\right)\epsilon^{-t/T_{dt}'} + \left(\frac{1}{X_{dt}''} - \frac{1}{X_{dt}'}\right)\epsilon^{-t/T_{dt}''}\right] \tag{8.44}$$

Fig. 8.10 (full line) shows the variation of the terminal voltage with time. There is a sudden drop from the original voltage U to the initial voltage $X_e U/(X_d'' + X_e)$ after the change. This initial voltage could be calculated alternatively by considering the generator to have a voltage U behind its subtransient reactance.

If the rapidly decaying subtransient component, given by the third term of Eqn. (8.44), is neglected the voltage drops suddenly from the original voltage U to the initial voltage $X_e U/(X_d' + X_e)$, which would be calculated by considering the generator to have a voltage U behind its transient reactance.

Finally the steady voltage $X_e U/(X_d + X_e)$ could be calculated in accordance with ordinary steady-state theory by considering the generator to have a voltage U behind its synchronous reactance.

Thus the curve showing the variation of the terminal voltage could be calculated by using the phasor diagram of Fig. 8.7 or Fig. 8.8 to determine the initial transient values and the steady value, and then fitting in appropriate time constants for the components of the change.

The transient time constant T_{dt}' depends on the external reactance. It is equal to T_d' when $X_e = 0$ (short-circuit), and is equal to T_{d0}' when $X_e = \infty$ (open circuit). For any given load impedance, T_{dt}' is intermediate between these two extremes. Similarly the subtransient time constant T_{dt}'' is intermediate between T_d'' and T_{d0}''.

A calculation of this kind is valuable in studying the action of a voltage regulator used to maintain a constant voltage. The regulator cannot act quickly enough to prevent the rapid drop to the initial transient voltage $X_e U/(X_d' + X_e)$, but it can restore the voltage to the original value U in a short time by automatically increasing the excitation, as indicated by the dotted line on Fig. 8.10. Such a curve can be calculated approximately by treating the action of the voltage regulator independently and using the principle of superposition. For an accurate calculation a step-by-step solution is necessary.

8.6.2 Rise of voltage after removal of load

During a load rejection test, in which the load is suddenly disconnected from a generator, the voltage rises, at first rapidly and then more slowly, until a new steady value is reached. The curve of variation of the voltage can be calculated by the method just described. Generally the subtransient effect is too rapid to be of any importance, and it is neglected in the following calculation.

If the phasor diagram for the original load condition is that shown in Fig. 8.7, the voltage behind synchronous reactance is U_0. Hence if saturation is neglected, the steady voltage after the load is removed is U_0. The voltage behind transient reactance is U', and is the initial value of the voltage immediately after the load is removed. The appropriate time constant is the open-circuit direct-axis transient time constant T_{d0}'. Hence the r.m.s. voltage is given by:

$$U_t = U' + (U_0 - U')(1 - \epsilon^{-t/T_{d0}'}) \tag{8.45}$$

and is plotted on Fig. 8.11. The full line includes also the subtransient component.

The practical problem is usually to find how rapidly a voltage regulator can restore the voltage to its normal value, and what maximum value is attained. The action of the regulator, as well as the effect of changing speed of the generator after disconnection from the system, can be studied approximately by using the principal of superposition. The dotted line in Fig. 8.11 shows how

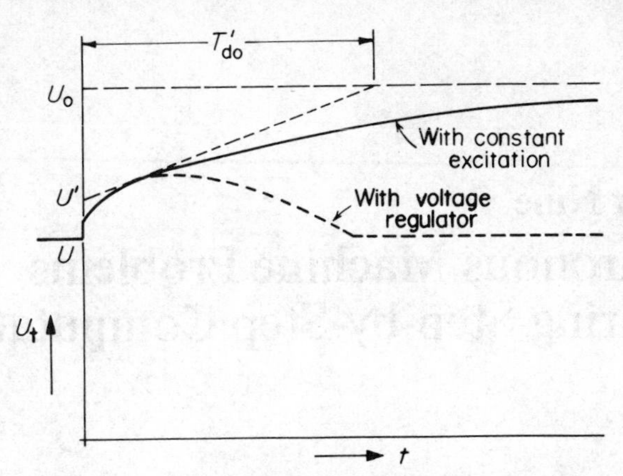

Fig. 8.11 Variation of the voltage after sudden removal of a load.

the voltage varies if a regulator is used. Although the value of the steady voltage U_0 determined by this method would be much too high because of saturation, the regulator limits its rise to a lower value. In practice the saturation is appreciable and, moreover, the speed rises. For an accurate calculation a step-by-step computation, discussed in Section 5.6, preferably allowing for saturation by the method indicated on p. 217 is necessary [36].

Chapter Nine
Synchronous Machine Problems Requiring Step-by-Step Computations

9.1 Transient stability

A synchronous generator connected to an infinite bus runs stably at synchronous speed and at a constant rotor angle if the mechanical input power from the prime mover exactly balances the generator losses and electrical output power, and if the conditions for steady-state stability are met. Any disturbance to the balance results in a change of speed and rotor angle. The generator is transiently stable only if after some temporary fluctuations the speed returns to synchronism and the rotor angle to a constant value. In the unstable condition the speed does not return to synchronism and the rotor angle consequently varies continuously. The calculation of the swing curve of rotor angle as a function of time, is therefore the prime objective of any transient stability study. This chapter studies the effect of a symmetrical three-phase short-circuit cleared after a short period of time. Although it occurs less often, the three-phase fault is particularly chosen since its effect is more severe than that of an unbalanced fault or any other disturbance.

The differential equation of motion of the machine is Eqn. (5.2), in which the electrical torque M_e depends on the flux and current variables. The simplified methods mentioned on p. 192, which do not allow in any case for regulating devices, are inadequate. For an accurate assessment of the transient stability of

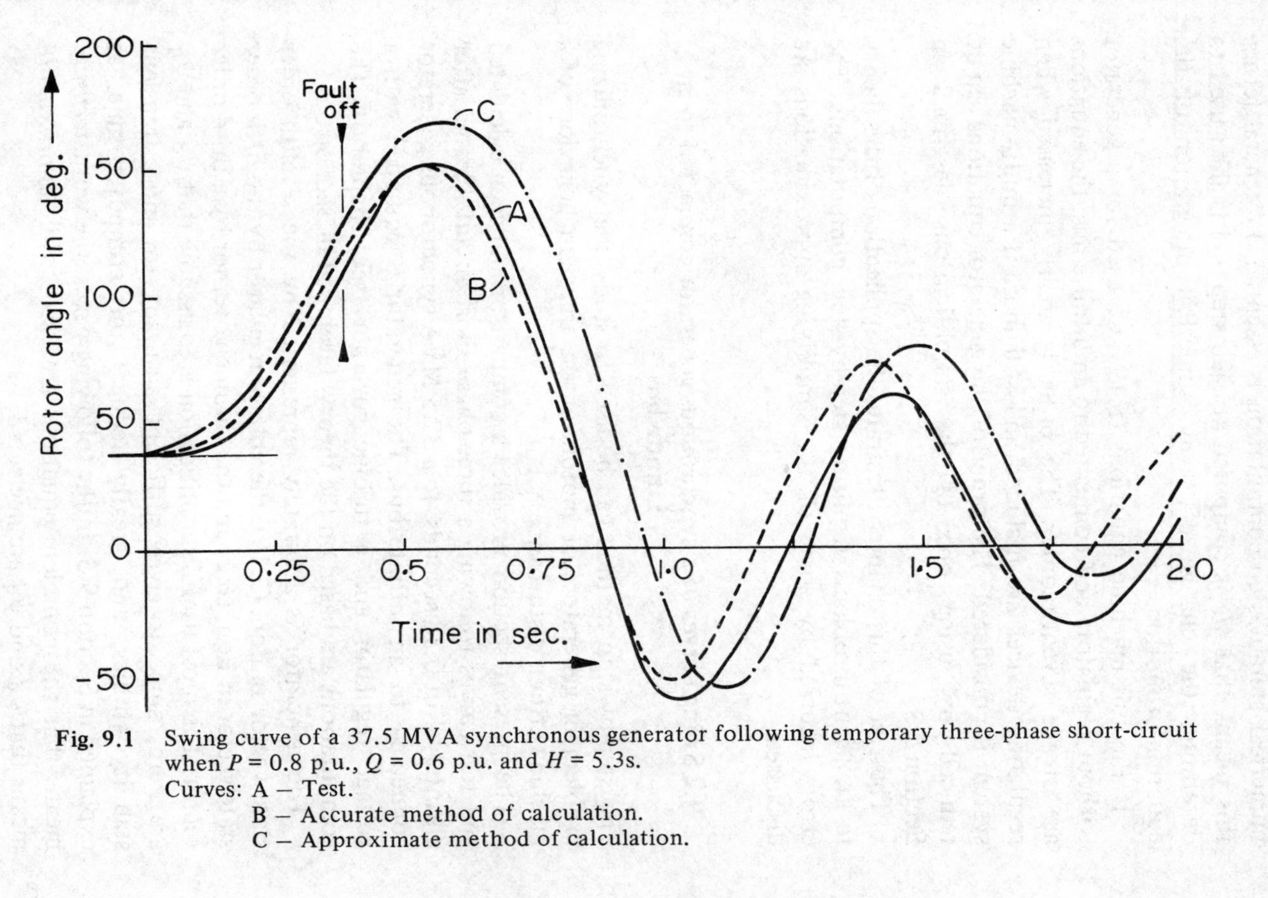

Fig. 9.1 Swing curve of a 37.5 MVA synchronous generator following temporary three-phase short-circuit when $P = 0.8$ p.u., $Q = 0.6$ p.u. and $H = 5.3$s.
Curves: A – Test.
B – Accurate method of calculation.
C – Approximate method of calculation.

a generator with a voltage regulator and a turbine governor, a numerical step-by-step computation is essential. In general terms, this means that M_e is computed at each step and is then used to determine the change of δ by numerical integration of the equation of motion.

If the full machine equations (5.19) are used for a generator without regulators and connected to an infinite bus, the equations are of the seventh order. The order is greatly increased when regulating devices are included and still more if a multi-machine system is considered. The order of the generator equations can be reduced by using one of the simplifications described in Section 5.3.

The rest of this chapter illustrates the application of the theory to transient problems requiring step-by-step computations. The errors incurred by some of the simplifying approximations are discussed.

9.2 Swing curves of a synchronous generator connected to an infinite bus

The step-by-step method can be used to study the synchronizing problem in generators or motors or the transient behaviour of a machine after a disturbance.

The present section provides a comparison between calculated and test curves following a three-phase short-circuit, cleared after an interval of 0.38 seconds, to a 37.5 MVA synchronous generator connected to an infinite bus. The generator is equipped with a voltage regulator and a turbine governor (see Chapter 6). The computations are based on the theory developed in Section 5.3.

The generator is accurately represented by the seventh order state vector in Eqn. (5.19), the voltage regulator by the fifth order state vector in Eqn. (6.2) and the turbine governor by a third order vector. The most useful simplification is to assume that $\dot{\psi}_d$ and $\dot{\psi}_q$ are zero, since it gives sufficient accuracy for most transient stability studies, and greatly reduces the computer time, as explained in Section 9.5. In the following pages, the word *accurate* means that the complete equations are used, while *approximate* means that $\dot{\psi}_d$ and $\dot{\psi}_q$ are neglected.

Fig. 9.1 shows the swing curves determined by a site test and by the accurate and approximate methods of calculation. The accurate curve shows good agreement with the test curve, especially during the period of the first swing, but the accuracy of curve C is probably adequate. There is more discrepancy during later swings, probably because of uncertainty in the turbine and governor parameters.

Fig. 9.2 shows computed curves of torque, flux linkage and current. They show clearly the oscillating components, which are present in the accurate solution, but disappear when $\dot{\psi}_d$ and $\dot{\psi}_q$ are neglected. The curves support the theoretical results of Sections 8.1 and 8.5, for which the additional assumption of constant speed was made.

Back-swing phenomenon

The losses occasioned by the heavy oscillating currents are associated with a nett braking torque and may in some cases be large enough to retard the rotor temporarily before acceleration commences. The temporary decrease in rotor angle is known as a *back swing* in the rotor position with respect to a synchronously rotating reference frame. There is no noticeable back swing in Fig. 9.1, although the approximate curve clearly indicates a more rapid increase in rotor angle during the fault period.

The back swing is more noticeable when the generator has little inertia and is lightly loaded, especially when it is over excited and connected to the infinite bus by a high reactance transmission line. The results of calculations, based on the same 37.5 MVA generator system, with a few modifications to the parameters to illustrate these effects, appear in Fig. 9.3.

The approximate solution does not contain the oscillating components and consequently does not show the back-swing effect. With a transient disturbance of a less severe type, for example, a sudden change of load, field voltage or external reactance, the error is less, so that, in view of the great saving in computer time, the approximate method is the one most commonly used. For further details the reader is referred to [29, 49, and 50].

9.3 Loss of synchronism of a synchronous generator. Effect on rectifier excitation systems

9.3.1 Loss of synchronism

A synchronous generator which loses synchronism but remains connected to the supply system may settle down after a transient

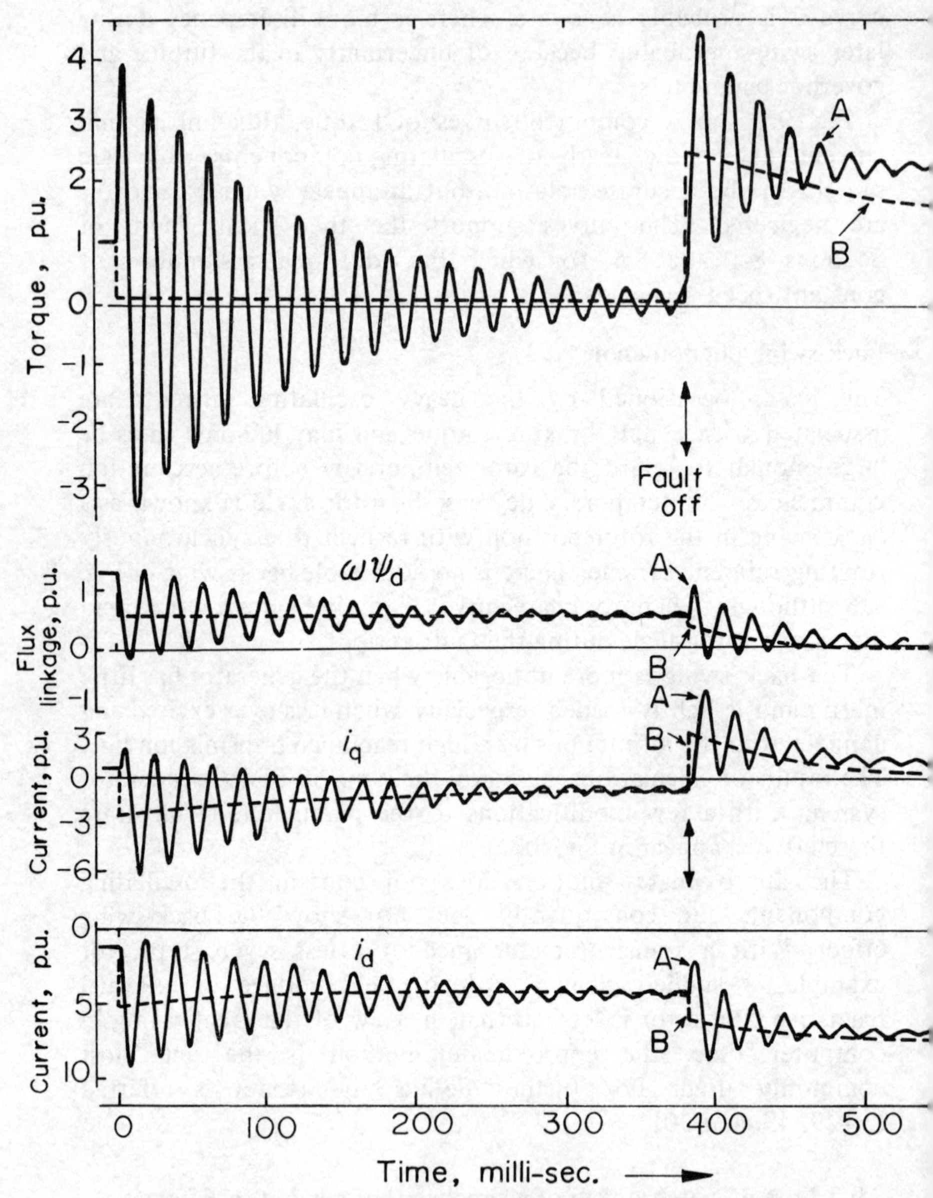

Fig. 9.2 Additional calculated results for the machine of Fig. 9.1.
Curves: A – Accurate Method.
B – Approximate Method.

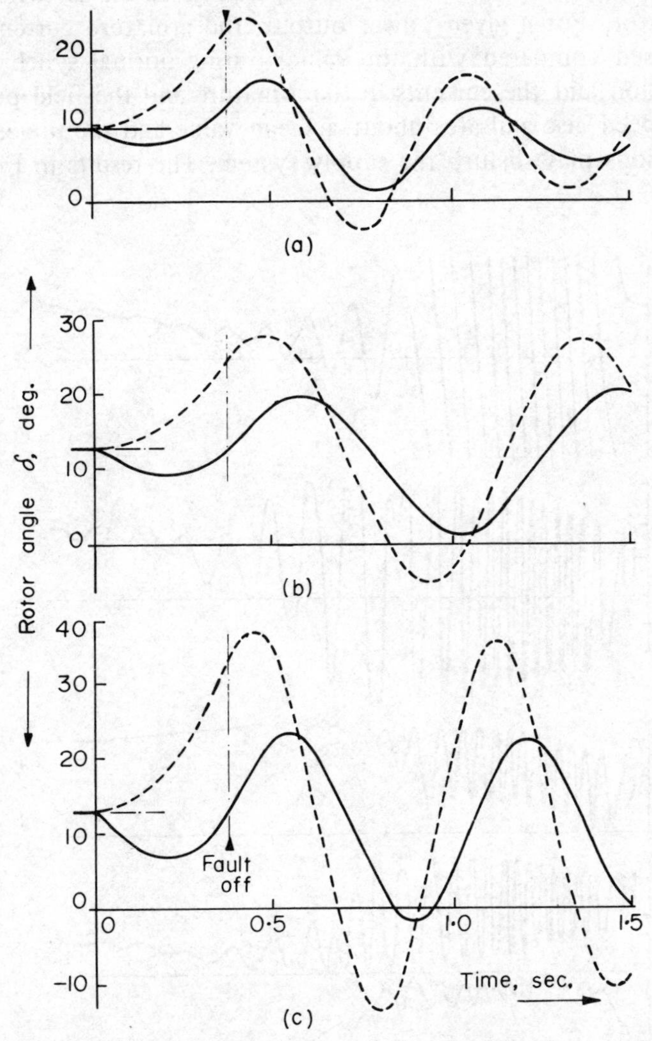

Fig. 9.3 Calculated swing curves of the 37.5 MVA synchronous generator following a temporary three-phase short-circuit when $P = 0.2$ p.u. and $Q = 0.6$ p.u.
——— accurate method; – – – approximate method.
(a) Tie-line reactance equals 0.18 p.u. and $H = 5.3$s
(b) Tie-line reactance equals 0.60 p.u. and $H = 5.3$s
(c) Tie-line reactance equals 0.60 p.u. and $H = 3.0$s

period to a steady condition of operation as an asynchronous generator. For a given power output, the armature currents are increased compared with the values during normal synchronous operation and the currents in the armature and the field pulsate. The speed also pulsates about a mean value and the undesirable pulsations may disturb the supply system. The results in Fig. 9.4

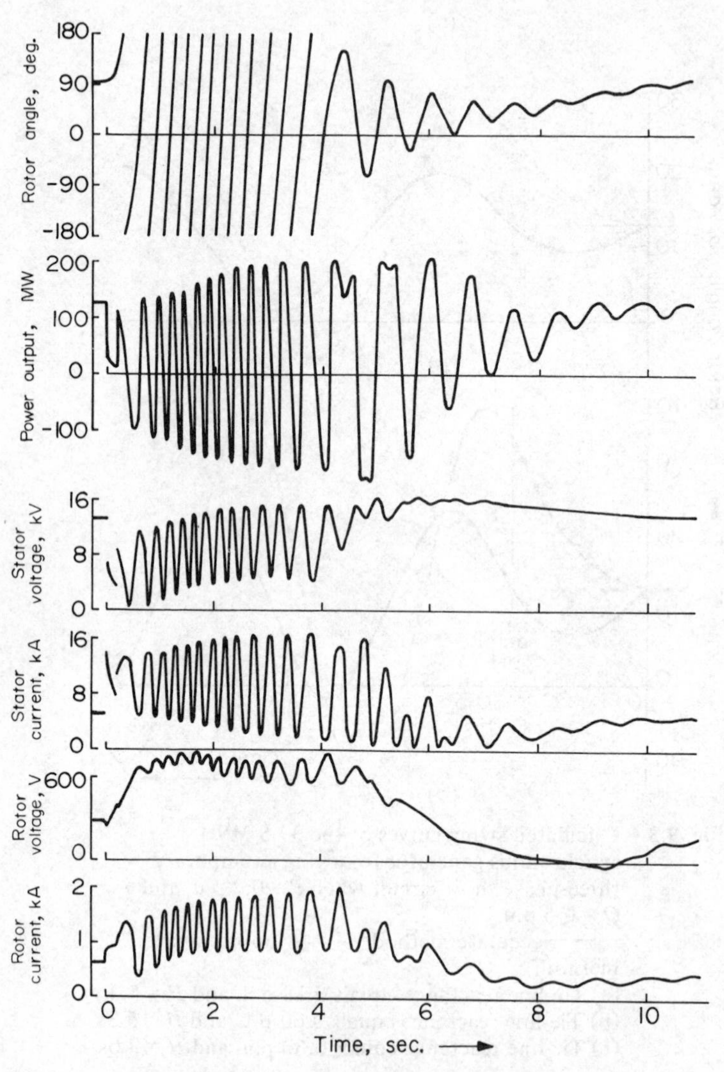

Fig. 9.4 Test results of a three-phase short circuit applied for 195 msec. to a 776 MVA synchronous generator.

illustrate the pulsations for the case of a 776 MVA turbo-generator which slipped 13 pairs of poles before resynchronizing [61].

It has been generally accepted in the past that a synchronous generator which loses synchronism must be disconnected forthwith. Recent experience supported by full scale tests has, however, indicated that under certain conditions and for a limited period it is permissible to leave the machine connected to the supply until some action can be taken to cause it to resynchronize. If, for example, the field excitation is removed when pole slipping starts, pulsations are much less severe. The field circuit would be closed, usually through a resistance. In any event, without excitation the machine becomes an induction generator with unsymmetrical secondary circuits. For further details of the practical considerations the reader is referred to [60 and 61].

During the transient period, when it is either just losing or regaining synchronism, the generator is accurately represented by the seventh order state vector in Eqn. (5.19). During steady asynchronous operation, the theory developed in Section 7.2 can be used, although the assumption that the speed is constant is not strictly true.

9.3.2 Effect on rectifier excitation systems

Large alternating voltages are induced in the field winding during asynchronous running and under normal conditions the field current reverses its polarity. In a rectifier excitation system the field current cannot change direction, because of the blocking action of the rectifier, and large inverse voltages, well in excess of the d.c. ceiling voltage of the excitation system, may appear across the rectifier elements. The results of a test [60] when blocking occurred on a 150 MVA generator appear in Fig. 9.5.

During the blocking period the field is effectively open circuited and the excitation source isolated; the field current is zero and the machine equations are modified. The machine response is calculated by a step-by-step numerical solution of the modified equations, allowing for the discontinuity in field current when blocking occurs. Two sets of state variable equations are therefore needed; one for the normal forward direction of field current and a second for the reverse direction or blocking condition. The step-by-step solution is transferred from one set to the other whenever the blocking either commences or ends [69].

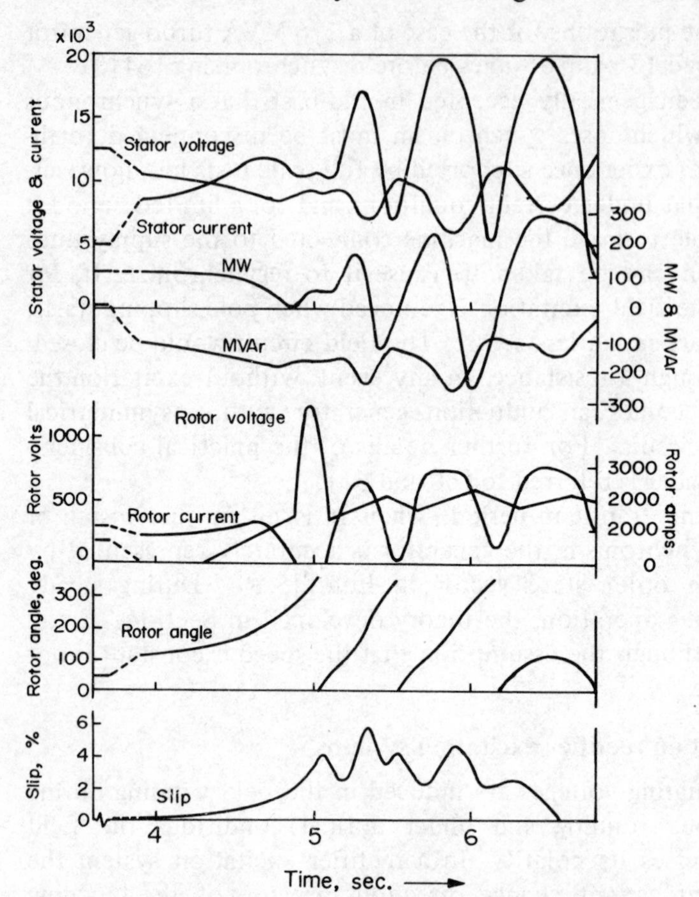

Fig. 9.5 Pole-slipping and restoration of synchronism of a 150 MVA synchronous generator with rectifier excitation.

9.4 Optimization of control inputs

The practical problem of determining the controls to optimize the performance of a system during a transient disturbance is a complicated one. Two examples of possible methods of dealing with it, using the methods explained in Section 5.5, are outlined below. For simplicity a system, consisting of a single machine connected to an infinite bus and to which Eqns. (5.19) apply, is considered.

Assuming that the generator is operating at a known steady condition, a transient disturbance is specified. The disturbance may, for example, be a sudden short-circuit which is cleared after a given time interval. A performance index, which serves as a measure of the quality of the performance, is selected. In [55] the performance index is

$$V = \int_{t_0}^{t_f} [A_1(x_1 - x_{1F})^2 + A_2(x_3 - x_{3F})^2 + A_3(z_1 - z_{1F})^2 + A_4(z_2 - z_{2F})^2]\,dt$$

A_1, etc. are weighting factors expressing the importance of the various state and control variables, of which the rotor angle $x_1 = \delta$ is the most important. The control variables must not exceed certain limiting values, expressing for example the ceiling voltage of the exciter or the limits of movement of the turbine valve gear. It is then required to determine the optimum control functions z_1 and z_2, which make V a minimum.

[55] evaluates the functions z_1 and z_2 by using Bellman's Eqn. (5.35) in conjunction with a second order dynamic programming method. The non-linear machine equations, with or without the simplifications discussed in Section 5.3, are solved to give z_1 and z_2 as functions of time [64].

The computation is lengthy even for a simple system and there is the disadvantage that the required control cannot readily be obtained by a fixed closed-loop feedback. To enable the control to be effective over a range of operating conditions, a different feedback transfer function is necessary for each initial operating point. The ideal solution would be to use an on-line computer to provide an adaptive control to meet this requirement.

An alternative proposal [70] bases the optimization on the linearized equations relating to the final steady operating condition. The calculation, which uses Riccati's equation (5.42), is then much simpler, but it is not necessarily fully applicable to the real non-linear system. It can however be checked by carrying out a forward computation to determine what response is obtained when the controls, calculated in this way, are applied to the non-linear system. The control functions for the linearized system are determined in a form which can be realized by feedback loops.

A further feature of the proposed method introduces a Liapunov function as a constraint on the optimization process in order to ensure that the resulting non-linear system, comprising the machine and its controls, is stable. After having found any Liapunov function which surrounds the operating point in the state-space diagram, the computation determines the control functions which force the trajectory inside the boundary. Clearly there is scope for important developments in this field.

9.5 Techniques for a multi-machine system

This chapter has so far described the single machine synchronized to an infinite bus, but in most practical cases it is necessary to consider a multi-machine system consisting of a number of machines connected together. Fig. 9.6 shows an example in which there is an infinite bus at one point of the system. Each machine is represented by the state variable equations in Section 5.3 together with the necessary equations for the excitation regulators and turbine governor.

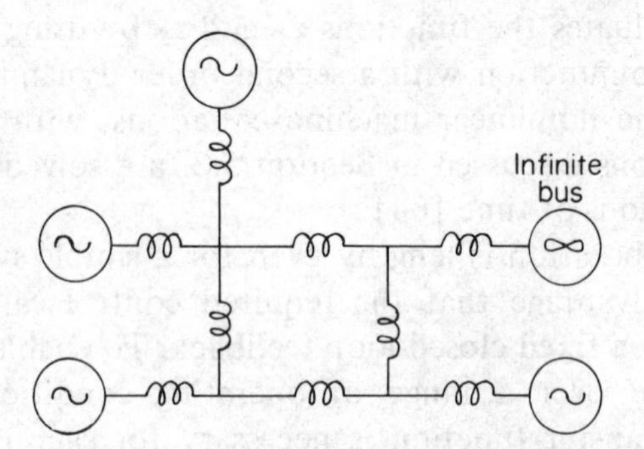

Fig. 9.6 Multi-machine system including an infinite bus.

The two-axis equations of Chapter 5 express the machine voltages and currents in a reference frame fixed to its direct and quadrature axes. However, when more than one machine is involved, the machines do not move together, each having its own rotor angle, and it becomes necessary to refer all quantities to a *common reference frame.* The preferred reference is the (D,Q) frame, rotating at synchronous speed since the transformed variables are the components of the phasor representing the armature variables (see Section 13.1). For each machine the variables in the two reference frames are related by a *frame transformation.*

It has been explained in Section 9.2 that the computation is considerably simplified and computation time reduced by neglecting the $\dot{\psi}_d$ and $\dot{\psi}_q$ terms in the machine equations. A similar advantage is gained by neglecting the $\dot{I}$ terms in the transmission line equations [40]. The frame transformation is unaffected by the above simplifications. In any machine the angle between the axis reference frame (d,q) and the synchronous reference frame (D,Q) is the rotor angle δ relative to the infinite bus. δ is an important variable because it is the mechanical angle which determines the inertia torque in the equation of motion. The transformation for the ith machine in an m-machine system is

$$(u_{\mathrm{D}i} - ju_{\mathrm{Q}i}) = (u_{\mathrm{d}i} - ju_{\mathrm{q}i})\epsilon^{-j\delta_i} \tag{9.1}$$

or

$$U_{\mathrm{DQ}i} = U_{\mathrm{dq}i}\epsilon^{-j\delta_i}$$

The current transformation is

$$I_{\mathrm{DQ}i} = I_{\mathrm{dq}i}\epsilon^{-j\delta_i}$$

Thus the transformation for any one machine is independent of those for all the other machines. The network equation relating the set of voltages U_{DQ} and currents I_{DQ} are

$$U_{\mathrm{DQ}} = Z_{\mathrm{DQ}}I_{\mathrm{DQ}} + L_{\mathrm{DQ}}\dot{I}_{\mathrm{DQ}} \tag{9.2}$$

where Z_{DQ} is the matrix of self and transfer complex impedances of the system and L_{DQ} is the matrix of network inductances.

Fig. 9.7 is a phasor diagram to illustrate, at any general operating condition, the position of the network voltage U in the common reference frame (D,Q) for either of two individual machines having reference frames (d_1,q_1) and (d_2,q_2). The angles by which the two rotors lag behind the synchronously rotating reference frame are δ_1 and δ_2.

9.5.1 Solution neglecting $\dot{\psi}_d$, $\dot{\psi}_q$, and $\dot{I}_{\mathrm{DQ}}$

With this simplification the axis currents and voltages in a machine are slowly changing quantities and the three-phase armature currents, and hence also the network currents, may therefore be represented by slowly changing phasors. The term $L_{\mathrm{DQ}}\dot{I}_{\mathrm{DQ}}$ in Eqn. (9.2) is zero and the network equations at any instant are algebraic phasor equations in complex numbers. Any transient solution depends on initial conditions as a starting point. At the

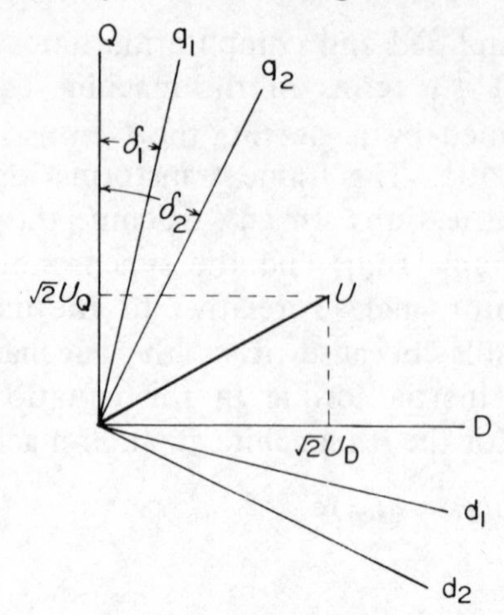

Fig. 9.7 Phasor diagram.

beginning of a swing curve calculation the steady initial values of network voltages and currents at the nodes connected to machines are found by a load-flow computation. The steady rotor angle of each machine is also determined. After any subsequent disturbance, like the clearance of the fault, the initial values are determined by the preceding calculation.

The transient response is found by a step-by-step numerical integration of the state equations of the system, which can be classified into three parts.

(a) Differential equations of the machines.
(b) Transformation equations.
(c) Algebraic equations of the network.

One step of the step-by-step calculation may be summarized as follows, starting with the machine state variables calculated by the previous step.

1. The axis currents i_d, i_q, of each machine at instant t_n are transformed by Eqn. (9.1) to i_D, i_Q, in the synchronous reference frame, using the known value of δ.
2. All the machine currents in the (D,Q) frame are used in an a.c. network calculation to determine the machine voltages u_D, u_Q.
3. The voltages u_D, u_Q, of each machine are transformed to axis voltages u_d, u_q, using Eqn. (9.1).

4. For each machine the state variable equations, with the known values of the variables at instant t_n, are used, in conjunction with the voltages u_d, u_q, to calculate the derivatives of the variables at instant t_n, and hence to find all the state variables at instant t_{n+1}.

There is clearly an advantage in dealing with the machine and network equations separately as in the above method, compared with a calculation using the very large number of equations required by a state variable formulation of the complete system.

9.5.2 Solution not neglecting $\dot{\psi}_d$, $\dot{\psi}_q$, and $\dot{I}_{DQ}$

As already explained with reference to Fig. 9.2, a solution obtained when $\dot{\psi}_d$, $\dot{\psi}_q$ are included in the machine equations, contains oscillations of the axis currents at supply frequency. The computer time is greatly increased, not only because each machine of a multi-machine system has two more state variable equations, but still more because the time step of the computation must be greatly reduced. If however, it is still permissible to neglect $\dot{I}_{DQ}$, Eqn. (9.2) is algebraic and the procedure of alternating between machines and network can still be used.

However, an operating condition which requires $\dot{\psi}_d$, $\dot{\psi}_q$ to be included in the machine equations would also require a better representation of the network. If $\dot{I}_{DQ}$ is included, its value cannot be computed directly from the network equations. There are two alternatives. One is to use a complete set of state variable equations for the whole multi-machine system. The other is to use the computing procedure described above and to determine $\dot{I}_{DQ}$ at each step by iteration. With either method the computer time is greatly increased.

The need to introduce $\dot{\psi}_d$, $\dot{\psi}_q$ arises when an accurate calculation of the back swing after a severe fault is required. (Section 9.2). The need also increases when the slip of the machine has a high value, for example, in an induction motor. (Section 11.4). The introduction of derivatives in the network equations becomes more necessary if they have to allow for capacitance in the transmission system. However, the simpler computation has so far been adequate for most practical conditions.

Some results of tests using the above methods are described in

Chapter 11 and shown in Fig. 11.4 for a two-machine system in which one machine was a micro-synchronous generator and the other a micro-induction motor.

9.5.3 Multi-machine system without an infinite bus

When it is an acceptable assumption to regard one point in the system as having a fixed supply of constant voltage and frequency – a so-called infinite bus – the computation is carried out as explained above. Although each machine swings relative to the infinite bus, the network frequency is constant.

A severe fault may however cause the whole system to drop in frequency, in which case no point can be taken as a fixed reference. It is then necessary to introduce a fictitious reference frame, usually the synchronous (D,Q) reference frame which would exist if the system continued to run at the original frequency without any disturbance. The rotor angle of each machine is then the position relative to the fictitious reference and the computation proceeds as before. If however, the drop in frequency is too great, the increase in slip of the machines may make it necessary to use the more accurate network representation of Eqn. (9.2) and moreover to reduce the reactance values in proportion to the frequency.

Chapter Ten
Effects of Saturation and Eddy Currents on Machine Performance

10.1 General

10.1.1 Saturation

The theories developed in Chapter 4 depend on the assumption that the magnetic material is unsaturated and that all fluxes are proportional to the currents producing them. Using the principle of superposition the equations are deduced by first finding the effect of individual currents and then superimposing all the effects. Although saturation is in practice very important, the equations obtained by neglecting it can often be used in performance calculations by determining effective constant parameters applicable to the particular problem.

The availability of the computer brings the hope that the effect of saturation can be allowed for more accurately. The difficulties now arise, not from the computation itself, but in the formulation of equations from which computations are made. Further discussion is given in Section 10.2.

10.1.2 Eddy currents

The assumption is made in Chapter 4 that the electrical conductors are of small section, over which the current density is uniform. It is also assumed that the flux in the magnetic circuit can be divided into a relatively small number of components (main and leakage) and that no currents flow in the magnetic material.

The eddy currents which flow when these assumptions are not true, can be considered under two headings, according to whether they occur in the magnetic material or in the electrical conductors.

Any magnetic material carrying alternating flux during normal operation must be laminated in order to limit the losses due to eddy currents. The calculation of such losses is a subject in itself and is to a great extent empirical, since they include not only the losses in the laminated material due to the main flux, but also losses due to high frequency pulsations at the air-gap surface and to stray fluxes linking the structural material of the machine. The total loss when the machine is excited but carries no load current is known as the *core loss*. It is a small proportion of the rated output but it has an important effect on the efficiency and the heating of the machine. For the purpose of making performance calculations it is usually either represented by a single resistance parameter or is neglected entirely. The core loss is an unavoidable defect of the machine and the design is carried out so as to make it as small as possible.

In a loaded machine the currents in the windings cause a redistribution of the flux, which is assumed to consist of separate main and leakage components. The core loss is thereby increased by an amount, called the *load loss*, which depends approximately on the square of the load current. To allow for this loss in performance calculations of synchronous machines, the armature resistance is increased. In induction motors a separate value of loss is deducted from the output. The effect may be important for motor starting, during which the *lost torque* is appreciable because of the high current.

Under normal steady conditions the flux in the secondary member does not vary with time in a synchronous machine and is of low frequency in an induction motor. Although there seems little need to laminate the secondary iron material, it is nevertheless common practice to laminate the entire magnetic circuit of an induction motor because of the pulsation losses mentioned above. In a synchronous machine there is usually some unlaminated iron in the secondary magnetic circuit.

The presence of unlaminated material, usually referred to as *solid iron*, in the secondary magnetic circuit has an important effect during any transient condition when the secondary flux varies. Often the eddy currents have a beneficial effect on the

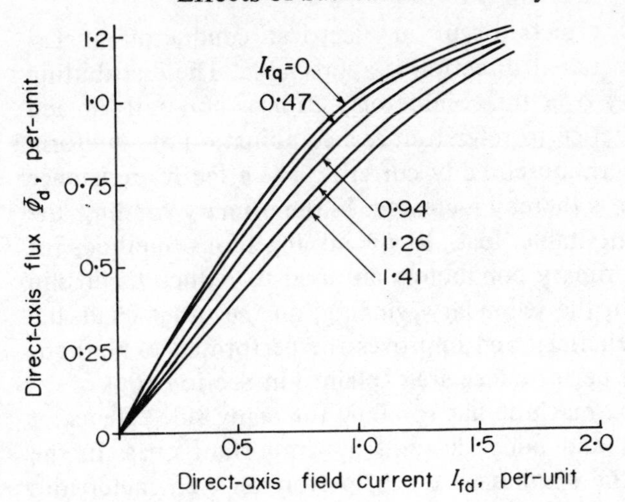

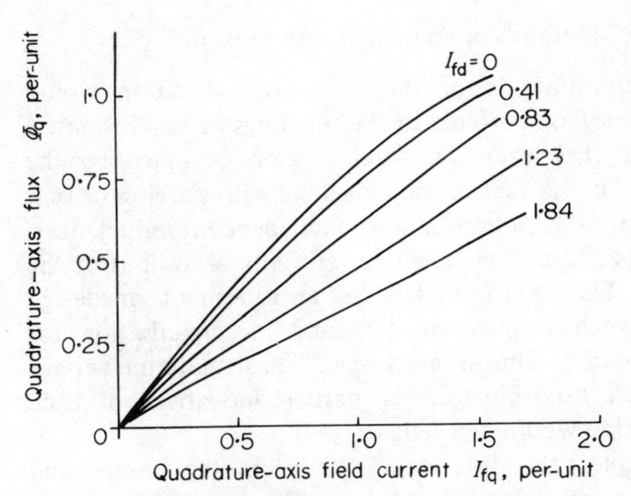

Fig. 10.1 The magnetisation curves of a small synchronous machine.

performance of the machine. Induction motors are sometimes built with a secondary member consisting of an iron cylinder, the eddy currents in which produce the motor torque. In a synchronous machine the eddy currents in solid material on the secondary member can provide the starting torque of a synchronous motor or the damping effect of a synchronous generator. Methods of calculating the effect on the performance of the machine are explained in Section 10.3.

Eddy current effects occur in electrical conductors if the dimension in the radial direction is appreciable. The distribution of current density over the conductor section is non-uniform and it is common practice to refer to it as a combination of a uniform current with superimposed eddy currents. The effective resistance of the conductor is thereby increased. In the primary winding, the result is an undesirable load loss, and ingenious methods of transposing the primary conductors are used to reduce the loss in large machines. In the secondary winding, on the other hand, the result is often beneficial and improves the performance. Methods of calculating the performance are explained in Section 10.4.

The design of a machine has to allow for many side effects, of which saturation and eddy currents are important ones. In the present chapter the emphasis is on the performance characteristics rather than on design details.

10.2 Methods of allowing for saturation

A complete determination of the effect of saturation would require, for any set of currents in the windings at any instant, a mapping of the flux over the whole region occupied by the machine. The six by six matrix for a machine with six circuits (c.f. Fig. 1.3 and Eqn. 4.1) contains nineteen independent inductances, each of which is a function of all six currents as well as of the rotor position. The determination by measurement or design calculation of such a mass of data and its organisation for computation would be almost impossible. To calculate the voltage induced by each flux linkage, six partial derivatives of each inductance function would be needed.

As an example, the interaction between the direct and quadrature-axis magnetizing inductances is illustrated by Fig. 10.1, which shows two families of curves measured on a small synchronous machine which had an additional field winding on the quadrature axis [21]. Each family shows, for each field winding, the magnetization curves obtained with various fixed values of current in the other winding. The curves were used to predict the steady-state performance, assuming all leakage inductances to be constant, and gave better agreement with test results than earlier methods. However, there was a great increase in complication even to allow for interaction between only two inductances.

A well-known method of dealing with a system having a limited non-linearity assumes that only one parameter is variable and that it varies with only one current. The method can be illustrated by considering the steady equivalent circuit of an induction motor (c.f. Fig. 3.4). The separation of the flux into a main flux and two leakage fluxes means that no differences of large quantities occur in the calculation. It can be assumed that the magnetizing reactance X_m is a function only of the magnetizing current, while the leakage reactances are constant. Such a method has been applied successfully to the calculation of the dynamic braking characteristics of an induction motor. Kingsley's method of 'saturated reactance' for synchronous machines is based on similar assumptions [7]. Many of the proposals [36] for allowing for saturation in calculations of transient performance of synchronous machines make similar assumptions. Sometimes it is assumed without any real justification that the same 'saturation factor' can be used for the direct and quadrature-axis reactances. The true situation is that both reactances depend on both currents.

The above method works quite well if the currents do not rise to high values, but is inaccurate when the currents are high enough to saturate the leakage flux paths. It is well known that the leakage reactance of an induction motor during starting at full voltage is appreciably less than during normal running. Since the main and leakage fluxes use the same iron, the effective magnetizing inductance of a synchronous machine at a given field current is greatly reduced when the armature current and hence the armature leakage flux increases to a high value. Thus a direct attack on the problem of saturation using the circuit equations, is not feasible, and thereby justifies the procedure of basing the general theory on the assumption of linearity. Depending on the problem, parameters based on total or incremental values of flux and derived from test or design curves, may be adequate. Often useful results can be obtained by minor modification of the linear theory which, although not rigorously valid, can at least give a result better than that obtained from the linear equations.

However, it has been shown by Erdelyi [62] and Silvester [53] that with the aid of a large computer it is possible in principle to obtain a more accurate solution by computing the flux distribution in detail. Erdelyi uses square or trapezoidal elements based on cartesian or polar co-ordinates while Silvester uses triangular elements. It is easier to fit triangular elements to the complicated boundaries in a machine but there is the disadvantage that the coefficients differ for each finite difference equation whereas they are the same for a regular square pattern. The number of equations is large, but it has been demonstrated in the papers referred to that such a calculation is quite feasible for steady operating conditions.

The method has been extended, using a triangular mesh, by Hannalla and MacDonald [71] to deal with transient problems. The time variation requires a step-by-step computation, and the field equations have to be solved at each step. The governing partial differential equation is

$$\frac{\partial}{\partial x}\left(\nu \frac{\partial A}{\partial x}\right) + \frac{\partial}{\partial y}\left(\nu \frac{\partial A}{\partial y}\right) = -J,$$

where A is the vector potential, which for a two-dimensional field system is a scalar function of x and y, and also varies with time.

Flux density and electric field strength are given by

$$B_x = \frac{\partial A}{\partial y}$$

$$B_y = -\frac{\partial A}{\partial x}$$

$$E = -\frac{\partial A}{\partial t}$$

J is the current density, which depends on the applied voltages and on the induced electric field. $\nu = 1/\mu$ is the local reluctivity, which is assumed to be a single-valued function of B.

The current density in any triangular mesh may be determined in two different ways.

1. In a continuous conducting medium, in particular the solid rotor iron or the distributed copper of a damper winding, $J = E/\rho$, where ρ is the resistivity.
2. In a winding consisting of conductors in series, which are of small section with negligible eddy currents, the current is determined by summating the electric field strength over the area occupied by the conductors, assuming them to be distributed over the slot area.

The above method can be applied directly to the field winding but there would be great complication in applying it exactly to an armature winding, which is in motion relative to the field. The following approximate method is proposed. Assume that the field structure is stationary (c.f. Fig. 1.3b) and that the armature structure moves relative to it at the instantaneous speed ω. Then replace the armature by a magnetic part which has infinite permeability and a uniform cylindrical surface with no slots. The three-phase armature winding is replaced by a current sheath at the surface having a sinusoidal distribution of linear current density. The current density distribution is defined completely by the two-axis currents i_d and i_q, which are calculated by summating the electric field strength in the conductors, having regard to the winding distribution, and the externally applied voltage. The circuit equations relating the axis voltages and currents would include a leakage inductance component using a value of inductance calculated by the normal method, and a zero-sequence current i_z can be introduced if the voltages are unbalanced. The air-gap length would be increased to allow for slot openings and for ampère-turns absorbed by the armature iron.

The approximation makes the same assumptions, principally in neglecting harmonic effects produced by the armature winding, as are made in the standard two-axis theory. The use of the two-axis transformation has the result that the field problem relates to a completely static medium. Probably the worst source of error would be that due to saturation in the armature teeth and core when large armature currents flow. Saturation and eddy currents in the iron carrying the field winding would be accurately allowed for, subject to the limited subdivision of the triangular meshes.

The field distribution at any instant and with given values of i_d, i_q and i_f, is computed by solving the finite difference equations. Because the reluctivity ν at any point is a function of the local flux density, an iterative procedure is necessary as in the calculation for a steady-state condition. To pass from one time step to the next a predictor method was found to be satisfactory, because any error in the initial prediction can be corrected in the course of the iteration carried out for the field computation.

Although the method, based on a determination of the field distribution at each time step, was first conceived as a means of allowing for saturation, it appears that for transient studies it is likely to be more important as a means of allowing for eddy currents.

10.3 Effects of eddy currents in the magnetic material

The overall distribution of flux density and current density in the complicated configuration of the magnetic material of the machine during a transient condition is exceedingly complex. If it can be assumed that no part of the iron is saturated, the magnetic conditions in the iron are linear although still complicated. Much of the theory which has been developed to calculate either core loss or damping effects, had been based on the assumption of constant permeability of the iron. It has also mainly dealt with conditions when all the quantities alternate at a given frequency, which presents less difficulty than the formulation of equations applicable to general transient conditions [12, 25].

Most texts on electrical machines assume that the flux distribution in a solid iron part of constant width, for example the yoke of a d.c. machine, is uniform and that the effect of the eddy currents in modifying this distribution under changing conditions may be neglected. However, eddy currents have the effect of concentrating the resultant flux and current near to the surface of the material – the so-called *skin effect.* Fig. 10.2 shows typical curves of the flux density distribution at different frequencies in an iron slab of finite width, with an alternating flux in the direction shown. At low frequencies the curve B differs little from uniform distribution (curve A). At high frequencies the flux density is high near the surfaces and low at the centre, as in curve C. The interpretation of the words 'high' and 'low' depends greatly on the thickness of the slab. Curve B would apply for 50 Hz in a thin lamination (0.5 mm), but 1 Hz in a turbo-generator rotor, for which the skin depth γ (see p. 225) is over 10 cm, would rank as a high frequency.

It is evident that at high frequencies the distribution near to either surface differs little from what it would be if the other surface were moved to infinity. Above a certain frequency it can be assumed that the distribution of flux, current and loss in a region just below a small element of the surface is the same as it would be if the element was part of an infinite plane surface

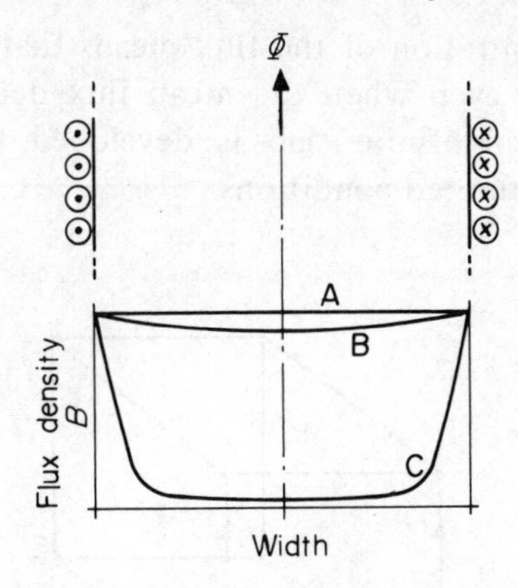

Fig. 10.2 Flux density distribution across the width of a solid iron slab at various frequencies.
Curves: (A) Uniform distribution.
(B) Distribution at low frequencies.
(C) Distribution at high frequencies.

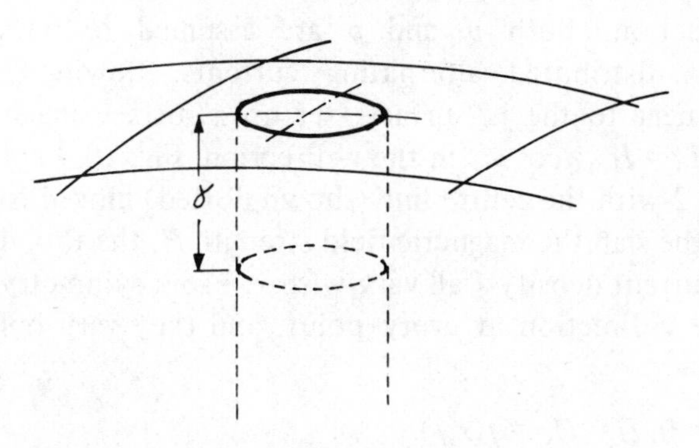

Fig. 10.3 Curved surface bounding a region of solid material.

bounding a semi-infinite slab. The same approximation can be made at any curved surface, as indicated by Fig. 10.3, if the effective skin depth γ is small compared with the radius of curvature.

Since the concentration of the flux means that saturation may become important even when the mean flux density is low, the theory of the semi-infinite slab is developed below for both saturated and unsaturated conditions.

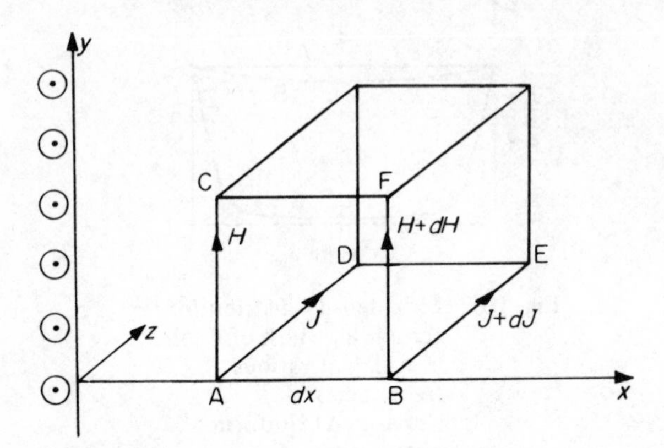

Fig. 10.4 Diagram of a semi-infinite slab and its co-ordinate axes.

10.3.1 Linear theory

Fig. 10.4 is a diagram of a semi-infinite slab of permeability μ and resistivity ρ, bounded by the yz-plane and extending to infinity in the x-direction. Both μ and ρ are assumed to be constant. Uniformly distributed alternating currents, flowing in the z-direction near to the yz-surface, set up a surface magnetic field strength $H_0 = H_{m0} \cos \omega t$ in the y-direction. Fig. 10.4 corresponds to Fig. 10.2 with the centre line (shown dotted) moved to infinity.

Inside the slab the magnetic field strength H, the flux density B, and the current density J all vary with x. From symmetry H and B are in the y-direction at every point, and they vary only with x and t. Thus

$$H_x = H_z = 0, H = H_y = f(x,t).$$

Also from symmetry

$$J_x = J_y = 0, J = J_z = F(x,t).$$

The electric field strength $E = \rho J$, is also in the z-direction. Applying Faraday's law to the surface element ADEB, in which AD is unit length

$$\left(E + \frac{\partial E}{\partial x} \cdot \mathrm{d}x\right)\mathrm{d}z - E \cdot \mathrm{d}z = \frac{\partial B}{\partial t}\,\mathrm{d}x\mathrm{d}z$$

or

$$\frac{\partial J}{\partial x} = \frac{1}{\rho} \cdot \frac{\partial B}{\partial t}\,. \tag{10.1}$$

Applying Ampere's law to the surface element ACFB, in which AC is unit length

$$J \cdot \mathrm{d}x \cdot \mathrm{d}y = \left(H + \frac{\partial H}{\partial x} \cdot \mathrm{d}x\right)\mathrm{d}y - H \cdot \mathrm{d}y$$

or

$$J = \frac{\partial H}{\partial x}\,. \tag{10.2}$$

Eqns. (10.1) and (10.2) are combined to yield

$$\frac{\partial^2 H}{\partial x^2} = \frac{1}{\rho}\,\frac{\partial B}{\partial t} \tag{10.3}$$

B and H are related by

$$B = \mu\mu_0 H \tag{10.4}$$

Combining Eqns. (10.3) and (10.4)

$$\frac{\partial^2 H}{\partial x^2} = \frac{\mu\mu_0}{\rho} \cdot \frac{\partial H}{\partial t} \tag{10.5}$$

For the particular case when the permeability μ is constant and the quantities alternate at angular frequency ω,

$$\frac{\mathrm{d}^2 \boldsymbol{H}}{\mathrm{d}x^2} = \frac{j\omega\mu\mu_0 \boldsymbol{H}}{\rho} = k\boldsymbol{H} \tag{10.6}$$

where $\boldsymbol{H}$ is the phasor representing H at the point x, so that

$$H = \mathrm{Re}(\boldsymbol{H}\epsilon^{j\omega t}) \tag{10.7}$$

The solution of Eqn. (10.6) is of the form

$$\boldsymbol{H} = A_1\epsilon^{kx} + A_2\epsilon^{-kx} \tag{10.8}$$

The constants A_1 and A_2 depend upon the boundary conditions

$\boldsymbol{H} = \boldsymbol{H}_0,\ x = 0$

$\boldsymbol{H} = 0,\ x = \infty.$

Hence

$H = H_0 \epsilon^{-(1+j)\alpha x}$

where H_0, the phasor representing the surface value $H_0 = H_{m0} \cos \omega t$, is equal to H_{m0}

$$\alpha = \sqrt{\frac{\omega \mu \mu_0}{2\rho}} \tag{10.9}$$

The flux density phasor is $B = B_0 \epsilon^{-(1+j)\alpha x}$

where $B_0 = \mu\mu_0 H_0 = \mu\mu_0 H_{m0} = B_{m0}$

The instantaneous value of flux density at any depth x from the surface is

$$\begin{aligned} B &= \mathrm{Re}[B\epsilon^{j\omega t}] \\ &= B_{m0}\epsilon^{-\alpha x}\cos(\omega t - \alpha x) \end{aligned} \tag{10.10}$$

The phasor representing the *total flux* through the slab, per unit *width* in the z-direction, is

$$\begin{aligned} \Phi &= \int_0^\infty B \, \mathrm{d}x \\ &= \frac{B_{m0}}{(1+j)\alpha} = \gamma B_{m0}\epsilon^{-j\pi/4} \end{aligned} \tag{10.11}$$

where $\gamma = 1/\alpha\sqrt{2} = \rho/\sqrt{\mu\mu_0}$, called the *skin depth* or *depth of penetration*, where depth is measured in the x-direction. The magnitude of the total flux is the same as if the flux density had the constant value B_{m0} for the depth γ.

Since H_0 depends on the excitation current and Φ determines the induced voltage, the effective impedance per unit width can be found. Eqn. (10.11) shows that the flux lags the excitation current by 45° and hence that the voltage leads the current by 45° [158].

10.3.2 Non-linear theory [28]

The full line curve OXA of Fig. 10.5a is the B–H curve of the material. The assumption of constant permeability is equivalent to replacing this curve by the straight line OA. A better approximation is obtained by using the horizontal straight line at a constant flux density B_s, where the value of B_s has to be

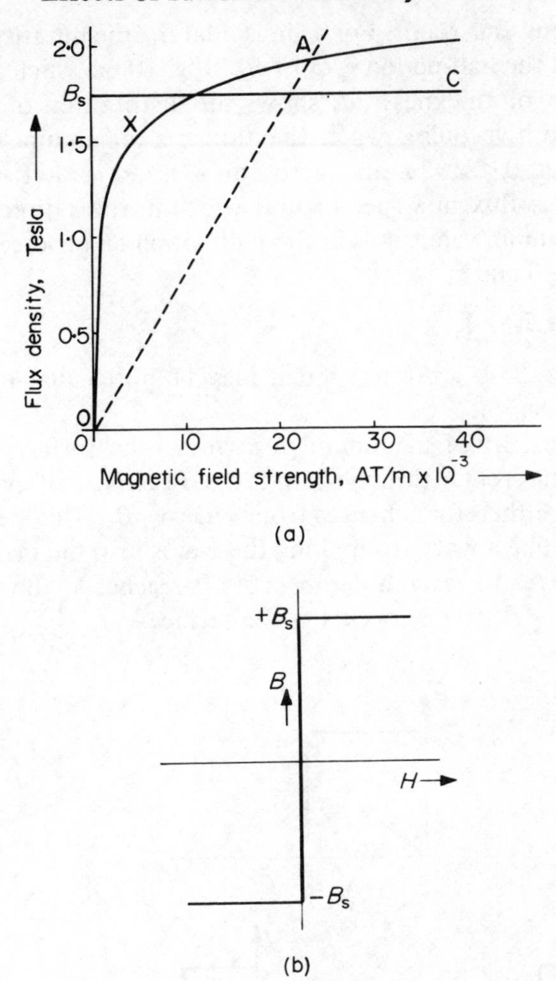

Fig. 10.5 Magnetization curves for forged steel.
(a) Curve OXA – Actual curve.
OA – Approximation with constant real permeability.
OB_sC – Non-linear approximation.
(b) Step curve.

determined empirically. In other words, the B–H curve is assumed to be a step-function as indicated in Fig. 10.5b.

For a cyclically varying surface magnetic field strength H_0, the surface flux density is a train of pulses, positive when H is positive and negative when H is negative, and these are propagated into the material along the x-axis. For a sinusoidal H_0 the duration of each pulse equals the half-period π/ω of H_0. Fig. 10.6a which applies to an iron slab of thickness $2d$, shows the distribution of B, at the end of a positive pulse, when the flux is a maximum. B has the value B_s for $0 < x < \delta$ and zero for $\delta < x < d$. Let Φ be the instantaneous flux in a slice of unit width in the z-direction, and Φ_m its maximum value. Φ is in the y-direction and is the total flux between $x = 0$ and $x = \delta$.

Hence $\Phi_m = B_s\delta$ (10.12)

At this stage B_s is arbitrary, but it may be noted that it must be greater than Φ_m/δ.

At the instant of maximum positive Φ, the surface magnetic field strength crosses through zero into a negative half-cycle and B at the surface therefore changes from $+B_s$ to $-B_s$. This step change in B moves like a wave front along the x-axis into the material and takes time π/ω to reach a depth δ. As it reaches δ, the next step change from $-B_s$ to $+B_s$ occurs at the surface.

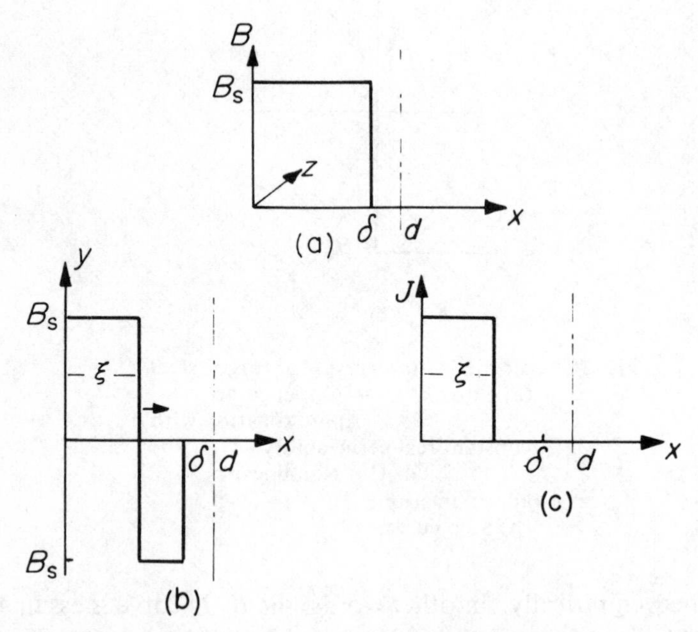

Fig. 10.6 Flux density and current distribution for non-linear theory.
(a) Flux density at instant of maximum flux.
(b) Flux density at any instant.
(c) Current density at any instant.

Fig. 10.6b shows the distribution of B during a positive half-cycle at the instant when the step change from $-B_s$ to $+B_s$ has just reached the point ξ. During this half-cycle, for a given value of ξ the current density J at depth x, as shown in Fig. 10.6c, depends on the electric field strength induced by the changing flux, and is given by

$$\left.\begin{aligned} J &= \frac{1}{\rho} \cdot \frac{\mathrm{d}}{\mathrm{d}t}\Big((\xi - x)B_s - (\delta - \xi)B_s\Big) \\ &= \frac{2B_s}{\rho} \cdot \frac{\mathrm{d}\xi}{\mathrm{d}t}, \qquad \text{for } x < \xi \\ &= 0, \qquad \text{for } x > \xi \end{aligned}\right\} \tag{10.13}$$

Note that

$\dfrac{\mathrm{d}x}{\mathrm{d}t} = 0$, because x is not a function of Time

Let zero time be the instant when the flux is a negative maximum, then

$$\begin{aligned} \xi &= 0 \text{ when } t = 0 \\ \xi &= \delta \text{ when } t = \pi/\omega \end{aligned} \tag{10.14}$$

At zero time Φ starts to change from its negative maximum value $-\Phi_m$, because H_0 is entering a positive half-cycle and B at the surface is changing from $-B_s$ to B_s. Hence

$$H_0 = H_{m0} \sin \omega t$$

Considering unit length in the y-direction, the surface value of H_0 (in the y-direction) is equal to the total current (in the z-direction) in the region $0 < x < \xi$. From Eqns. (10.12) and (10.13)

$$H_0 = H_{m0} \sin \omega t = \int_0^d J\mathrm{d}x = \frac{2B_s\xi}{\rho} \cdot \frac{\mathrm{d}\xi}{\mathrm{d}t} = \frac{B_s}{\rho} \cdot \frac{\mathrm{d}}{\mathrm{d}t}(\xi^2) \tag{10.15}$$

Integrating Eqn. (10.15) and using the boundary conditions of Eqn. (10.14)

$$\xi^2 = \frac{\rho H_{m0}}{\omega B_s}(1 - \cos \omega t)$$

and

$$\delta^2 = \frac{2\rho H_{m0}}{\omega B_s} \tag{10.16}$$

Thus

$$\xi = \delta \sin(\omega t/2) \tag{10.17}$$

over the period

$$0 < t < \pi/\omega$$

Since δ is the maximum value of ξ,

$$\Phi_m = B_s \delta = \sqrt{\frac{2H_{m0}B_s\rho}{\omega}}$$

Now from Fig. 10.6b, using Eqn. (10.17), the instantaneous value of total flux is

$$\Phi = B_s\xi - B_s(\delta - \xi)$$

$$= \Phi_m \left(\frac{2\xi}{\delta} - 1\right)$$

$$= \Phi_m (2 \sin(\omega t/2) - 1) \tag{10.18}$$

Curves showing the variation of Φ and H_0 with time are given in Fig. 10.7. Φ is a discontinuous curve with each half-cycle similar to the preceding one, but reversed in direction. The fundamental is

$$\Phi_1 = \frac{4\sqrt{5}}{3\pi} \Phi_{1m} \sin(\omega t - 63.4^0) \tag{10.19}$$

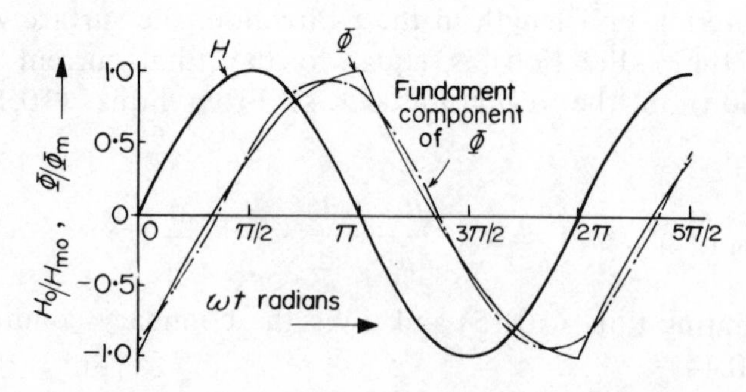

Fig. 10.7 Waves of magnetic field strength and flux based on the non-linear theory.

where

$$\Phi_{1m} = 2\omega\Phi_m = \sqrt{(8H_{m0}B_s\omega\rho)}$$

Φ_1 is the effective flux for unit axial length of the machine (direction z in Fig. 10.6) from which the impedance at any frequency can be found.

The result differs from that given by the linear theory in two important respects,

(a) The magnitude of Φ_{1m} varies as $\sqrt{H_{mo}}$ and so introduces a non-linear relationship.
(b) The angle of lag between the fundamental component of Φ and H is still constant, but is now 63.4° instead of 45°.

In order to apply the eddy current theory to practical problems, much judicious simplification of the conditions in the machine is necessary. The theories apply to conditions when the flux is alternating at a given frequency and can be used directly to determine the performance when the machine runs asynchronously at constant slip. They determine the damping action of a turbo-generator or the starting characteristics of a synchronous motor with solid iron in the rotor. The direct-axis equivalent circuit is shown in Fig. 10.8 which is the circuit of Fig. 7.2a with the damper branch replaced by the variable parameter Z_{kd}.

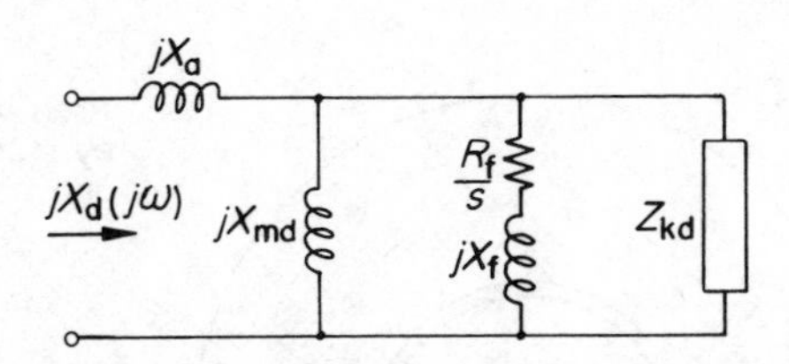

Fig. 10.8 Equivalent circuit giving $X_d(j\omega)$.

10.3.3 Application to turbo-generator damping

Any design method based on the theory must be to some extent empirical, since there are many uncertainties. Because the rotor is not infinitely long there are appreciable end-effects. Also the resistivity ρ of the iron depends on its temperature, which varies greatly during asynchronous operation. One suggestion is to use an empirical value of ρ, based on earlier test results, to allow for both the temperature and the end-effects.

A simple assumption led to a useful method of studying the asynchronous operation of a turbo-generator at low slip. Consider the condition when the rotor is at rest in a position where the d-axis coincides with the flux axis of two armature phases in series. The direct-axis flux, when the two stator phases are excited at low frequency by a single-phase voltage, is assumed to flow as indicated by the dotted lines in Fig. 10.9. Because of the skin effect caused by the eddy currents in the rotor iron, the flux is forced outwards and is concentrated in a band of iron material at each side of the rotor body. Since the area of the surface at which the skin effect occurs is known, the flux due to a given m.m.f. and hence the effective impedance can be calculated. Details of the calculation are given in [28]. Fig. 10.10 gives the frequency locus of the direct-axis operational impedance, with the field circuit open, by measurement and calculation for a 75 MVA turbo-generator. The experimental curve A is much flatter than the semicircle B, which would be obtained with a single discrete damper winding. Curve C, calculated by the non-linear theory, is much closer to the measured curve than curve D, calculated by the linear theory.

In addition to the end effects and variable resistivity, there are other sources of error in the theory.

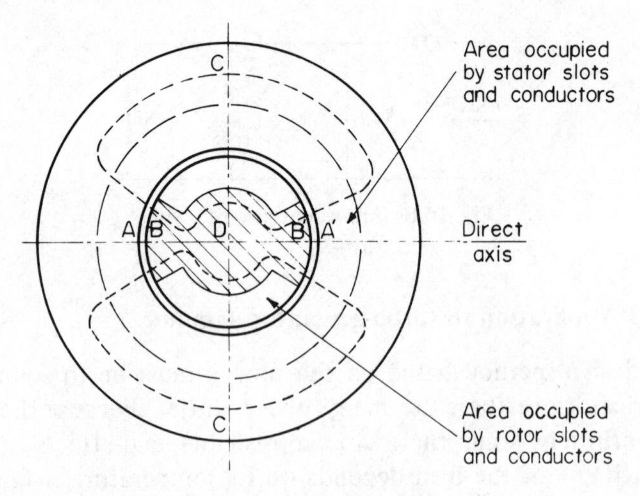

Fig. 10.9 Direct-axis flux path of a turbo-generator.

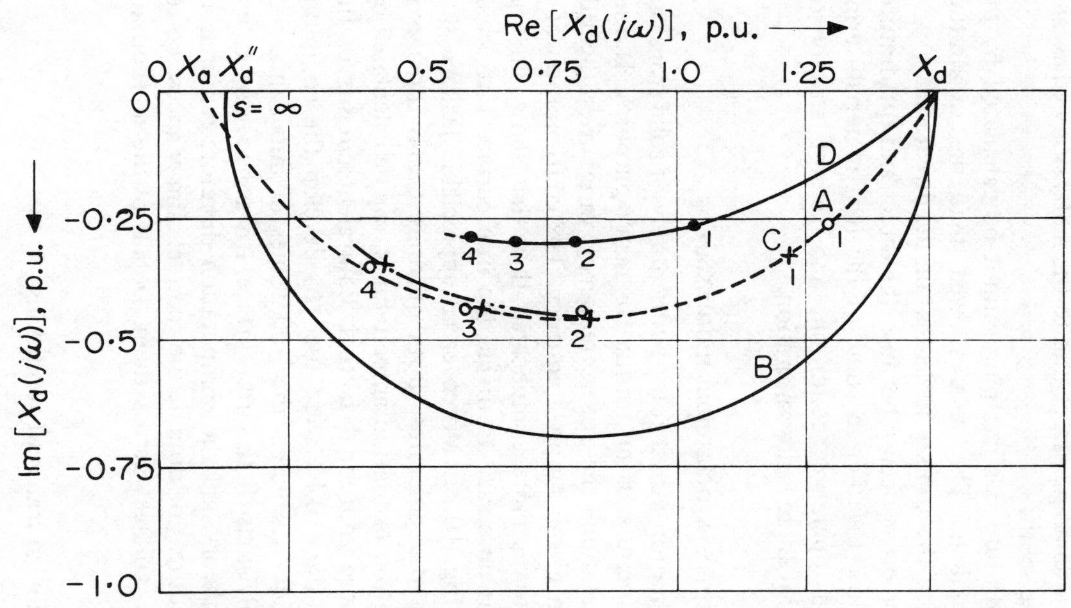

Fig. 10.10 Impedance locus with field open of 75 MVA generator
Curves: (A) (0 – 0) Experimental locus.
(B) Calculated for idealized machine.
(C) (x – x) Calculated by non-linear method.
(D) Calculated assuming constant real permeability.

The numbers indicate points at the same value of slip.

(a) The flux does not enter the rotor only in the limited region indicated as the skin depth. Some flux enters through the teeth and some is spread over the pole face in the central region, which is also subject to skin effect.
(b) There is a damping effect due to the metal slot wedges and the iron of the teeth.
(c) The quadrature-axis damping cannot be calculated by the method described. The tests showed that the quadrature damping was of the same order as that on the direct axis.
(d) A further error occurs because the direct and quadrature-axis fluxes flow together in some of the iron material during any practical operating condition, and cannot strictly be superimposed if there is any saturation.

10.3.4 Application to synchronous motor starting

A synchronous motor with a rotor built entirely of solid iron, can develop an adequate starting torque for many purposes. During the starting operation, the rotor frequency varies over the full range from zero to the supply frequency. The configuration of a salient-pole motor is more complicated than that shown in Fig. 10.9 and it was necessary to divide the iron surface into small sections, as in Fig. 10.11. Moreover an appreciable leakage flux passes between the pole tips. Both quadrature and direct-axis fluxes, due to given excitation ampere-turns, can be calculated by allowing for these additional factors. Equivalent circuits of the type shown in Fig. 10.12 were used to develop the computer programme which calculated the 100 values of impedance representing 100 divisions of the iron surface. It was necessary to use the non-linear theory and the computation therefore required an iteration procedure to obtain the correct impedances corresponding to the voltages across them. Details of the method are given in [41].

10.3.5 Application to transient conditions

Both the linear and non-linear theories given, were worked out for alternating conditions. If it is permissible to use the linear theory, the frequency response function $X_d(j\omega)$ can be converted into the Laplace transform $X_d(p)$ and the time response deduced by

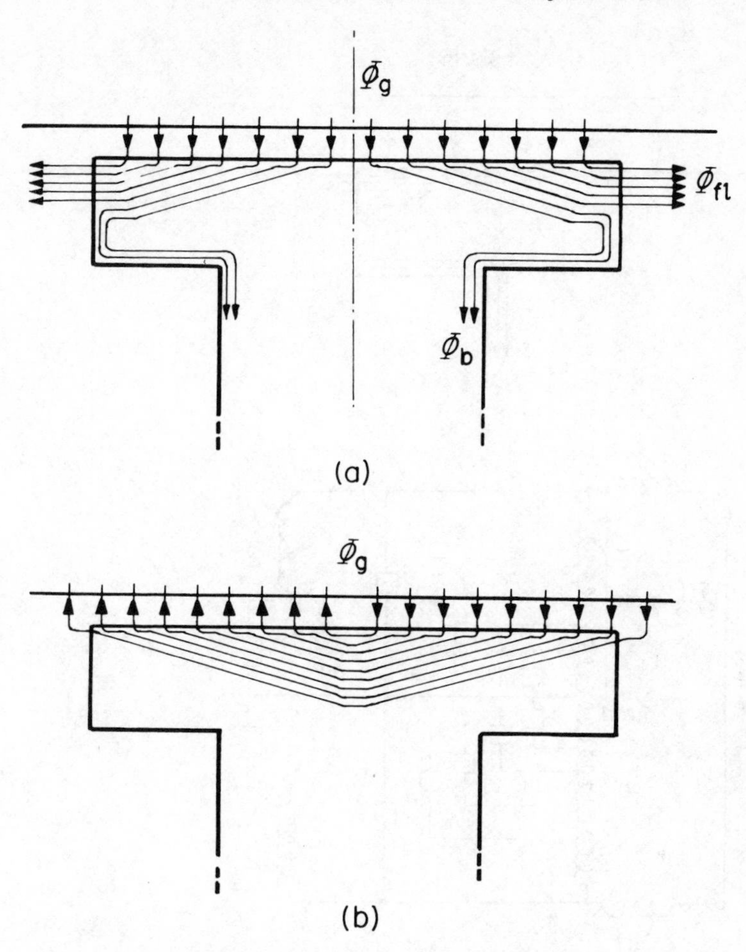

Fig. 10.11 Distribution of flux in the solid iron.
(a) Direct-axis flux. (b) Quadrature-axis flux.

inversion. The method is applied tentatively in [28] to a sudden short-circuit and at least explains why the subtransient component of the short-circuit current is in practice not an exponential curve as is predicted by the general theory which does not allow for eddy current effects. It is however not easy to apply the method to more complicated transient conditions and if it is necessary to use the non-linear theory, the only method that offers hope of an accurate result is that based on the electromagnetic field, described briefly on p. 219.

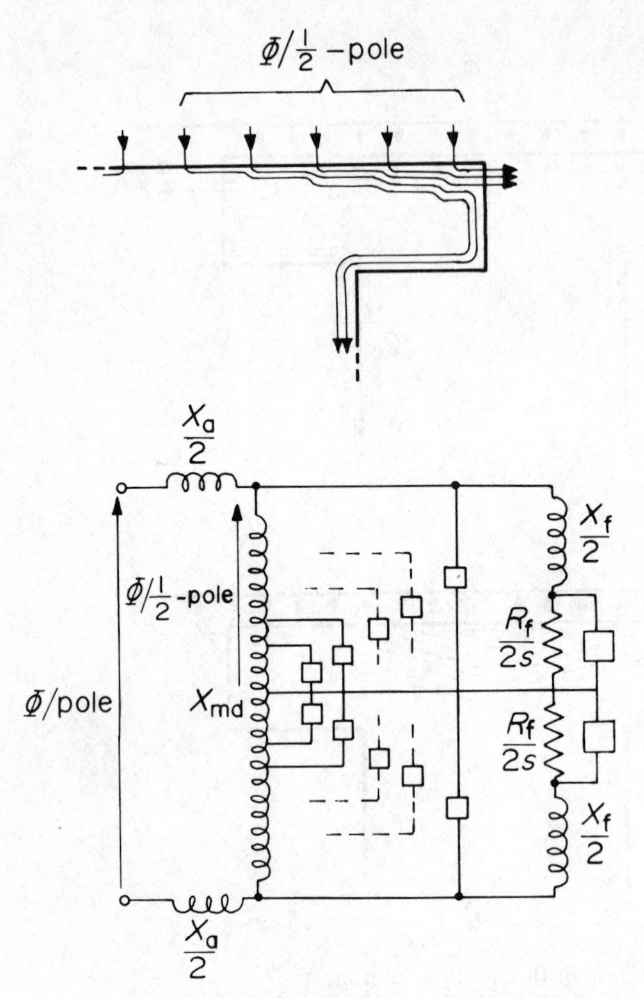

Fig. 10.12 Direct-axis flux paths and the representative equivalent circuit.

10.4 Effect of eddy currents in the rotor conductors

The theory of eddy currents in the machine conductors is easier to apply, mainly because, for this problem, it is justifiable to neglect saturation. It is usually assumed that the iron in the flux paths considered is infinitely permeable, and that the flux due to a given current depends only on the air space through which it passes. The problem considered in the next few pages is that of a cage-type induction motor, in which the rotor conductors are deep enough to cause the current density to vary appreciably over the section. As before the calculations are made for an alternating condition.

Consider first a double-cage induction motor, which has two separate cages at different radial distances from the air-gap. A simple (and well-known) modification of the steady-state theory leads to Eqn. (10.20) instead of Eqn. (3.3).

$$\left.\begin{aligned} U &= U_i + (R_1 + jX_1)I_1 \\ 0 &= sU_i + (R_2 + jsX_2)I_2 + jsX_{23}(I_2 + I_3) \\ 0 &= sU_i + jsX_{23}(I_2 + I_3) + (R_3 + jsX_3)I_3 \\ I_m &= I_1 + I_2 + I_3 \\ U_i &= jX_m I_m . \end{aligned}\right\} \qquad (10.20)$$

Fig. 10.13a gives the equivalent circuit. The suffix 3 indicates the third winding and X_{23} is the reactance corresponding to the leakage flux which links windings 2 and 3, but not winding 1. A deep-bar cage affects the motor characteristics in a similar way, but a different calculation is necessary. The equivalent circuit is Fig. 10.13b in which Z_b is a variable impedance. Typical slot shapes for deep-bar cages are shown in Fig. 10.14.

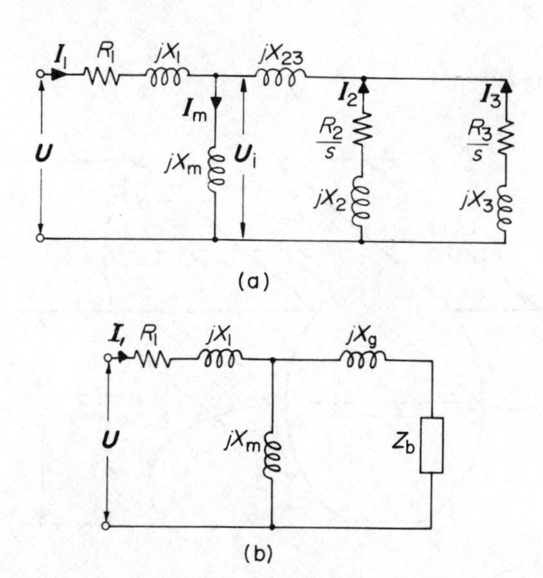

Fig. 10.13 Equivalent circuits.
(a) Double-cage induction motor.
(b) Deep-bar cage induction motor.

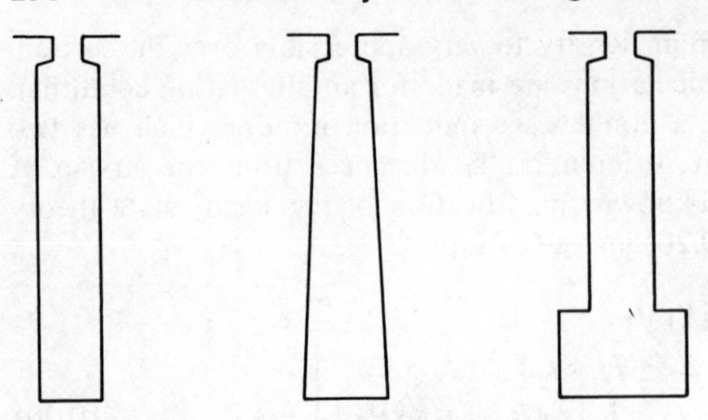

Fig. 10.14 Typical slot shapes for deep-bar cages.

Consider a slot of any shape, as illustrated in Fig. 10.15 in which $y = f(x)$. The slot is assumed to be full of non-magnetic material up to the level at which $x = h$, and it is assumed that the flux lines pass horizontally across the slot. The current density J is a function of x and t. From Ampere's Law, the flux density in the slice dx at level x depends on the total current below it. Hence

$$B = \mu_0 H = \frac{\mu_0}{y} \int_0^x Jy \, \mathrm{d}x$$

so that

$$\frac{\partial}{\partial x}(By) = \mu_0 Jy \tag{10.21}$$

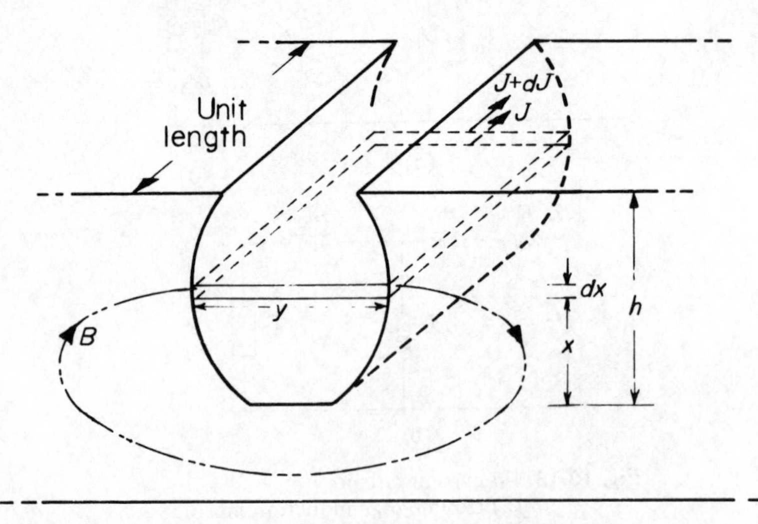

Fig. 10.15 Diagram of a slot filled with conducting material.

or

$$y\frac{\partial B}{\partial x} + B \cdot \frac{\mathrm{d}y}{\mathrm{d}x} = \mu_0 J y \tag{10.22}$$

Applying Faraday's law to the slice, the difference between the values of electric field strength E at the levels x and $(x + \mathrm{d}x)$ [compare Eqn. (10.1)] is

$$\mathrm{d}E = \rho\,\mathrm{d}J = \frac{\partial}{\partial t}(B\,\mathrm{d}x)$$

or

$$\frac{\partial J}{\partial x} = \frac{1}{\rho} \cdot \frac{\partial B}{\partial t}. \tag{10.23}$$

Differentiating Eqn. (10.22) with respect to t and using Eqn. (10.23)

$$\frac{\partial^2 J}{\partial x^2} + \frac{1}{y} \cdot \frac{\mathrm{d}y}{\mathrm{d}x} \cdot \frac{\partial J}{\partial x} - \frac{\mu_0}{\rho} \cdot \frac{\partial J}{\partial t} = 0. \tag{10.24}$$

For the particular case when the quantities alternate at angular frequency ω,

$$\frac{\mathrm{d}^2 \boldsymbol{J}}{\mathrm{d}x^2} + \frac{1}{y} \cdot \frac{\mathrm{d}y}{\mathrm{d}x} \cdot \frac{\mathrm{d}\boldsymbol{J}}{\mathrm{d}x} - \frac{j\omega\mu_0 \boldsymbol{J}}{\rho} = 0 \tag{10.25}$$

where $\boldsymbol{J}$ is the phasor representing J at point x so that

$$J = \mathrm{Re}\,[\boldsymbol{J}e^{j\omega t}]$$

Using the boundary conditions that

$$\boldsymbol{B} = \frac{\mathrm{d}\boldsymbol{J}}{\mathrm{d}x} = 0 \quad \text{when } x = 0$$

$$\boldsymbol{E} = \boldsymbol{E}_0 = \rho\boldsymbol{J} \quad \text{when } x = h$$

the effective impedance per unit length of bar in the z-direction is found from

$$Z_\mathrm{b} = \frac{\boldsymbol{E}_0}{\int_0^h \boldsymbol{J}y \cdot \mathrm{d}x} \tag{10.26}$$

A computer solution can readily be obtained for any shape of slot. For a rectangular slot, y is equal to the constant width b, and $\mathrm{d}y/\mathrm{d}x$ is zero. Eqn. (10.24) therefore reverts to the form of Eqn. (10.5) and

$$Z_b = \frac{1}{b}\sqrt{\frac{\omega\mu_0\rho}{2}} \cdot (1+j) \cdot \frac{\sinh 2\theta - j\sin 2\theta}{\cosh 2\theta - \cos 2\theta} \tag{10.27}$$

where

$$\theta = h\sqrt{\frac{\omega\mu_0}{2\rho}}.$$

Additional leakage reactance X_g must be added to the bar impedance, determined by evaluating Eqn. (10.26), to allow for the air space above the top of the metal and also for end-effects.

Curve A in Fig. 10.16 shows the calculated admittance locus of a deep-bar machine [56]. It was found that a curve of the type applicable to a double-cage machine could be fitted quite closely to it, as shown by curve B. Curve B for the double-cage, is obtained as the summation of values on two semi-circles of the type appropriate for a machine with two secondary windings, and relatively simple formulas are available for the parameters in the operational admittance function. Thus the method using an equivalent double-cage provides a means of calculating the performance of a deep-bar induction motor under both transient or alternating conditions. The method first determines the actual

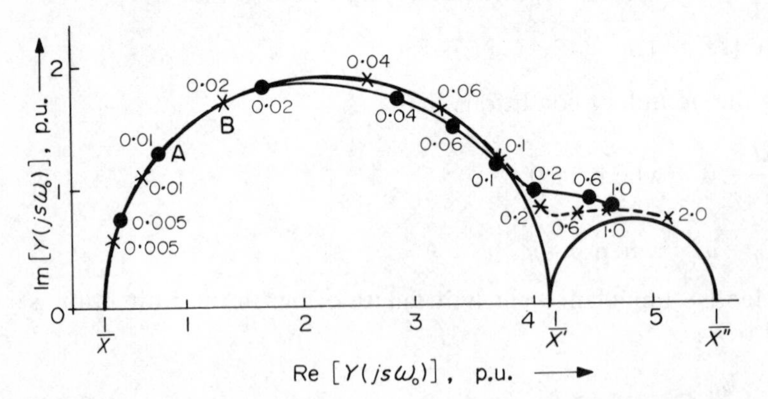

Fig. 10.16 Admittance locus. Curves: (A) (● — ●) Deep-bar cage.
(B) (x — x) Double-cage.

Figures denote value of slip.

frequency locus from Eqn. (10.26) by computer or otherwise, then fits the nearest double-cage type locus to it and finally uses the transient and subtransient parameters so obtained to calculate the performance under any transient condition using methods very similar to those in common use for synchronous machines. The subject is developed further in the next chapter.

Chapter Eleven
Induction Motor Problems

11.1 Application of equations in primary reference frame

It was explained in Section 4.8 that a two-axis representation of the induction motor can be made in two different ways. The more direct method, in which the reference frame is attached to the primary element (normally the stator), is useful for many purposes, particularly in making calculations for a machine with a single secondary winding. The equations are (4.46) and (4.47).

An important application is for studying unbalanced conditions, of which the capacitor motor, a common form of small single-phase motor, is a typical example. A stator quadrature winding Q_1 has a capacitor C in series and is connected to the same single-phase voltage as the direct-axis winding D_1, as in Fig. 11.1. The capacitive reactance is $Z = 1/j\omega_0 C$. Converting Eqns. (4.46) and (4.47) into phasor equations and making the necessary modifications, the two-axis steady state equations of the capacitor motor (assuming equal numbers of turns in D_1 and Q_1) are,

$$\begin{bmatrix} U \\ 0 \\ 0 \\ U \end{bmatrix} = \begin{bmatrix} R_1 + jX_1 & jX_m & & \\ jX_m & R_2 + jX_2 & (1-s)X_2 & (1-s)X_m \\ -(1-s)X_m & -(1-s)X_2 & R_2 + jX_2 & jX_m \\ & & jX_m & R_1 + jX_1 + Z \end{bmatrix} \begin{bmatrix} I_{d1} \\ I_{d2} \\ I_{q2} \\ I_{q1} \end{bmatrix} \qquad (11.1)$$

where $X_m = \omega_0 L_m$
$X_1 = \omega_0 L_{11}$
$X_2 = \omega_0 L_{22}$

The equations in the primary reference frame have also been used to study transient conditions in an induction motor [37]. However, such calculations are most often required for large cage-type machines where the rotor has deep bars and cannot be accurately represented by a single secondary winding. In such a case the analysis is simplified if the reference frame is attached to the secondary member [57].

11.2 Equations in secondary reference frame. Complex form of the equations

The equations of a machine with a single secondary winding are (4.44) and (4.45). They correspond to Park's equations for the synchronous machine, except that there is no field winding, the air gap is uniform, and all parameters are the same on both axes.

In the form which introduces the flux linkage variables, as in Eqns. (4.15) and (4.23), the voltage equations for an induction motor with two secondary windings are as follows. The zero sequence equation is omitted and the additional inductance L_{23} is included as in Eqn. (10.20).

$$\left.\begin{aligned} u_{d1} &= p\psi_{d1} + \omega\psi_{q1} + R_1 i_{d1} \\ u_{q1} &= -\omega\psi_{d1} + p\psi_{q1} + R_1 i_{q1} \end{aligned}\right\} \qquad (11.2)$$

$$\left.\begin{aligned} \psi_{d1} &= (L_m + L_1)i_{d1} + L_m i_{d2} + L_m i_{d3} \\ u_{d2} &= 0 = pL_m i_{d1} + [R_2 + p(L_m + L_2 + L_{23})]\, i_{d2} \\ &\quad + p(L_m + L_{23})i_{d3} \\ u_{d3} &= 0 = pL_m i_{d1} + p(L_m + L_{23})i_{d2} \\ &\quad + [R_3 + p(L_m + L_3 + L_{23})]\, i_{d3} \end{aligned}\right\} \qquad (11.3)$$

$$\left.\begin{aligned} \psi_{q1} &= (L_m + L_1)i_{q1} + L_m i_{q2} + L_m i_{q3} \\ u_{q2} &= 0 = pL_m i_{q1} + [R_2 + p(L_m + L_2 + L_{23})]\, i_{q2} \\ &\quad + p(L_m + L_{23})i_{q3} \\ u_{q3} &= 0 = pL_m i_{q1} + p(L_m + L_{23})i_{q2} \\ &\quad + [R_3 + p(L_m + L_3 + L_{23})]\, i_{q3} \end{aligned}\right\} \qquad (11.4)$$

Because of the symmetry of the magnetic system, Eqns. (11.2) to (11.4) can be replaced by half the number of complex equations, if new complex variables U_1, I_1, I_2, I_3, Ψ_1, are introduced. They are obtained by combining pairs of variables, e.g.,

$$U_1 = u_{d1} - ju_{q1} \tag{11.5}$$

$$I_1 = i_{d1} - ji_{q1} \tag{11.6}$$

The minus sign is chosen so that during steady operation the expression for U_1 agrees with that given by Eqn. (7.4). The equations become,

$$U_1 = p\Psi_1 + j\omega\Psi_1 + R_1 I_1 \tag{11.7}$$

$$\left.\begin{aligned}\Psi_1 &= (L_m + L_1)I_1 + L_m I_2 + L_m I_3\\ 0 &= pL_m I_1 + [R_2 + p(L_m + L_2 + L_{23})]I_2\\ &\quad + p(L_m + L_{23})L_3\\ 0 &= pL_m I_1 + p(L_m + L_{23})I_2 + [R_3 + p(L_m + L_3\\ &\quad + L_{23})]I_3\end{aligned}\right\} \tag{11.8}$$

It should be noted that the use of the complex number here is quite different from its use in converting sinusoidal quantities into phasors. It is here used in the normal mathematical method of combining two scalar variables into a single complex variable, often thereby simplifying the mathematics, particularly when

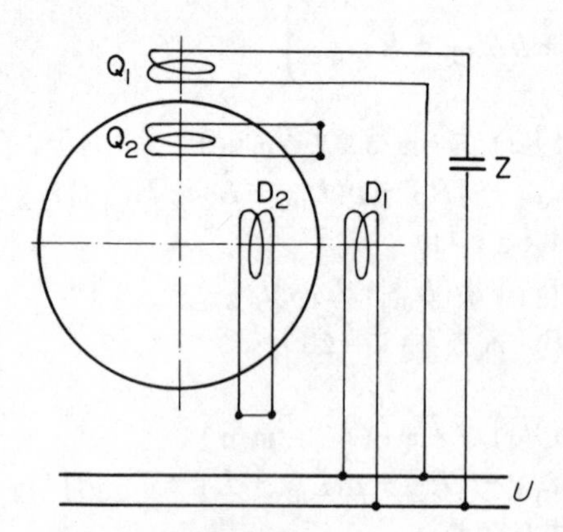

Fig. 11.1 Diagram of a single-phase capacitor type induction motor.

exponential and sinusoidal functions are involved. It is possible moreover to apply the complex equations to sinusoidal operation by substituting $p = j\omega$ and introducing phasors to represent the resultant sinusoidal quantities. Although the complex numbers are introduced in two different ways, the manipulation is still carried out by the normal methods of complex algebra.

Taking Laplace transforms and eliminating I_2 and I_3 from Eqns. (11.8) in a manner similar to that used for the synchronous machine, $\bar{\Psi}_1$ is determined as a function of $\bar{I}_1$.

$$\omega_0 \bar{\Psi}_1 = X(p)\bar{I}_1 \tag{11.9}$$

$X(p)$ is the operational impedance of the induction motor. It has the same value on each axis and, for the machine with two secondary windings, has the form,

$$X(p) = \frac{(1 + T'p)(1 + T''p)}{(1 + T_0'p)(1 + T_0''p)} X \tag{11.10}$$

where

$$X \quad = \omega_0 (L_m + L_1) \tag{11.11}$$

X is the 'synchronous reactance', equal to the sum of the magnetising reactance and the leakage reactance of the primary winding. The operational equivalent circuit includes the inductance L_{23}, as in Fig. 11.2. Because of the presence of L_{23} the formulas for the four time constants in Eqn. (11.10) are similar to those in Section 4.7. The form of the expression as the quotient of two quadratic functions is the same as in Eqn. (4.29) and transient and subtransient reactances X' and X'' can be determined from X and the time constants, as on p. 89.

11.3 Short-circuit and fault currents due to induction motors

Large induction motors of many MW are now in common use and it is important to be able to predict their effect on short-circuit or fault currents in the system. Many of the motors have deep-bar cage windings in order to limit the on-line starting current. As explained in Section 10.4, it is usually possible to replace the deep-bar cage by an equivalent double cage, so that the equations of the last section can be used for calculations. The method also has the advantage that the results correspond very closely to those obtained by the well-established methods explained in Section 8.5 for calculating the fault currents due to synchronous machines.

The induction motor consists of coupled circuits on the stator and rotor similar to those of the synchronous machine. The principal differences, none of which affects the application of the method, are

(a) The induction motor has no excitation winding and the secondary winding corresponds to the damper winding of the synchronous machine. A double-cage winding however requires two transient reactances and four transient time constants, which can conveniently be called transient and subtransient, although the winding arrangement differs somewhat from that of the synchronous machine.

(b) The induction motor normally operates with a small slip.

(c) When isolated from the supply, an induction motor, unlike a synchronous machine, cannot generate its full voltage for any appreciable time, because there is no excitation. Consequently the procedure for carrying out a sudden short-circuit test has to be slightly different.

(d) The induction motor is fully symmetrical between the direct and quadrature axes. Thus with similar approximations, the accuracy of the calculations should be better than for a synchronous machine.

A short-circuit test, made in order to determine the transient parameters, can be carried out in two ways. For a *direct short-circuit*, the primary winding terminals are suddenly shorted while they are connected to the supply, usually when the motor is unloaded with only the magnetizing current flowing. Since the supply is also shorted, there may be difficulties in making this test. An alternative is the *indirect short-circuit* test, for which the motor is first disconnected from the supply and then shorted after a short interval. Before the short-circuit, if friction is neglected, the unloaded motor runs at zero slip and for all the calculations of this section, it is assumed that the speed remains constant at synchronous speed. The problem therefore belongs to the third category in Table 5.1 on p. 99. The machine equations are linear and are solved by Laplace transforms, using the principle of superposition in a manner similar to that in Section 8.1. The notation used for the original steady quantities and for superimposed values is that explained on p. 39.

11.3.1 Direct short-circuit

The phase voltage before the short-circuit is assumed to be

$$u_a = U_m \sin(\omega_0 t + \lambda) \\ = \mathrm{Re}(\boldsymbol{U}_1 e^{j\omega_0 t}) \tag{11.12}$$

and

$$\boldsymbol{U}_1 = -jU_m e^{j\lambda} \tag{11.13}$$

$\boldsymbol{U}_1$ is the phasor representing this voltage alternating at frequency ω_0 and the angle λ defines the instant of fault application in the a.c. cycle. With the machine running at synchronous speed, the complex voltage U_1 is equal to the phasor $\boldsymbol{U}_1$. (see p. 134 and 264).

The steady no-load current I_{10} lags 90° behind the voltage, if core loss and armature resistance are neglected. The error in neglecting R_1 in Eqn. (11.14) when determining the no-load current is small, but it should be noted that it is not legitimate to neglect R_1 in Eqn. (11.15), because R_1 determines the time constant T_a, at which the asymmetrical components of the armature currents decay. As with the voltage, the complex axis current I_1 is equal to the phasor $\boldsymbol{I}_1$. Hence

$$I_{10} = \boldsymbol{I}_{10} = \frac{\boldsymbol{U}_1}{jX} = \frac{-U_m e^{j\lambda}}{X} = -_j I_m e^{j\lambda} \tag{11.14}$$

The relation between the superimposed voltage and current in the Laplace domain is found from Eqns. (11.7) and (11.9).

$$\bar{I}_1{}' = \frac{\omega_0 \bar{U}_1{}'}{X(p)\left(p + j\omega + \dfrac{R_1 \omega_0}{X(p)}\right)} \tag{11.15}$$

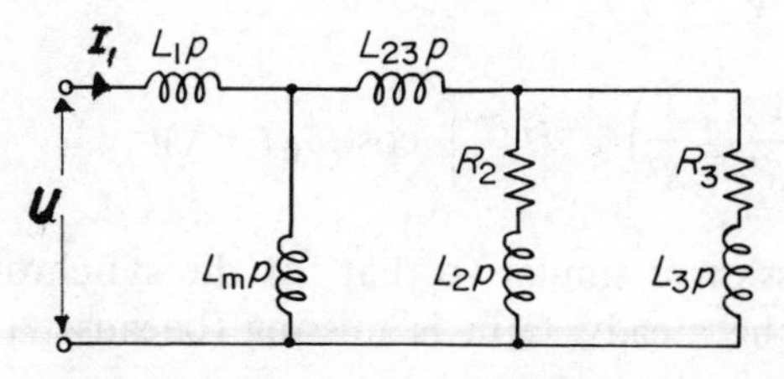

Fig. 11.2 Operational equivalent circuit of double-cage induction motor.

The voltage before the short-circuit is given by Eqn. (11.13). Since this is suddenly reduced to zero at $t = 0$, the superimposed voltage is

$$U_1' = jU_m \epsilon^{j\lambda} \tag{11.16}$$

and its Laplace transform is

$$\bar{U}_1' = \frac{jU_m \epsilon^{j\lambda}}{p} \tag{11.17}$$

Now the term $R_1 \omega_0 / X(p)$ in Eqn. (11.15) can be replaced as an approximation by a constant α, following the same argument as that used in Section 8.1 for the synchronous machine.

$$\alpha = \frac{R_1 \omega_0}{X} = \frac{1}{T_a} \tag{11.18}$$

Thus

$$\bar{I}_1' = \frac{j\omega_0 U_m \epsilon^{j\lambda}}{p(p + \alpha + j\omega_0)X(p)} \tag{11.19}$$

The inverse transform I_1' is obtained by putting $X(p)$ into partial fractions and using the inverse transformation Eqn. (13.14). After adding the original current I_{10} to obtain I_1, the instantaneous current i_a is found by

$$i_a = \mathrm{Re}(I_1 \epsilon^{j\omega_0 t}) = \mathrm{Re}(I_1 \epsilon^{j\omega_0 t}) \tag{11.20}$$

$$i_a = \frac{-U_m}{X''} \epsilon^{-t/T_a} \cos\lambda + \left[U_m \left(\frac{1}{X'} - \frac{1}{X} \right) \epsilon^{-t/T'} + U_m \left(\frac{1}{X''} - \frac{1}{X'} \right) \epsilon^{-t/T''} \right] \cos(\omega_0 t + \lambda) \tag{11.21}$$

The expression is similar to that for the synchronous machine, except that the steady term is missing (because I_{10} was added), and hence the current decays to zero.

For a short period after zero time, the alternating component can be considered to be the sinusoidal wave obtained by putting $\epsilon^{-t/T'}$ and $\epsilon^{-t/T''}$ equal to unity in the previous equation. Hence

$$i_{a(a.c.)} = U_m \left(\frac{1}{X''} - \frac{1}{X}\right) \cos(\omega_0 t + \lambda)$$

$$= U_m \operatorname{Re}\left[\left(\frac{1}{X''} - \frac{1}{X}\right) \epsilon^{j(\omega_0 t + \lambda)}\right]. \qquad (11.22)$$

Using Eqns. (11.14), (11.20) and (11.22), the phasor equation for the a.c. component of short-circuit current is found as

$$\boldsymbol{I}_1 = \frac{j}{X''}(\boldsymbol{U}_1 - j\boldsymbol{I}_{10}X'')$$

$$= \frac{j\boldsymbol{U}''}{X''}. \qquad (11.23)$$

For a short period after any sudden change the machine can therefore be represented by the voltage $\boldsymbol{U}''$ in series with the subtransient reactance X''. The value of I_1 calculated from Eqn. (11.23) corresponds to the initial part of the a.c. component. The result is similar to that deduced for the synchronous machine in Section 8.5.2.

11.3.2 Open-circuit voltage after disconnection

When the supply to the induction motor is interrupted, the flux linking the inductive windings does not allow the current to decrease instantaneously to zero, but an arc is drawn in the switch until the point of zero current is reached. An exact calculation for the three-phase system would be very complicated, and the method used here assumes that the decay of each current is specified by multiplying that current by a factor $\epsilon^{-\sigma t}$. The solution contains an initial voltage peak, which depends on σ, and would be an infinite impulse voltage if the current were reduced instantaneously to zero.

Before the open circuit the original voltage and current are given by Eqns. (11.13) and (11.14). The solution is obtained by adding the original voltage U_{10} to the change of voltage due to a sudden change of current

$$I_1{}' = I_{10}(1 - \epsilon^{-\sigma t})$$

applied to an initially dead system. The change in voltage is found in the Laplace domain from Eqn. (11.15) as

$$\bar{U}_1{}' = -I_{10}\,\frac{\sigma}{\omega_0}\cdot\frac{(p + a + j\omega_0)}{p(p + \sigma)}\,X(p). \tag{11.24}$$

The solution for the terminal voltage after the open circuit is

$$u_a = \left(\frac{-\sigma X''}{\omega_0}\,\epsilon^{-\sigma t}\,I_m\right)\sin(\omega_0 t + \lambda) + [(X - X')\epsilon^{-t/T_0{}'} + (X' - X'')\epsilon^{-t/T_0{}''}]I_m\cos(\omega_0 t + \lambda) \tag{11.25}$$

The first term in Eqn. (11.25) is a sharp peak at the origin. However, because of the arcing at the switch, the current interruption is delayed and the effect is not noticeable in practice. It is therefore neglected in the rest of this section. The second and third terms are the transient and subtransient alternating components respectively. However, after a few cycles when the short circuit is applied at the instant t_1, the subtransient component has become negligible and the voltage becomes

$$u_a = I_m(X - X')\epsilon^{-t/T_0{}'}\cos(\omega_0 t + \lambda) \tag{11.26}$$

For a short period after zero time the terminal voltage can be considered approximately to be the sinusoidal wave obtained by putting $\epsilon^{-t/T_0{}'}$ equal to unity in Eqn. (11.26).

Hence

$$\begin{aligned} u_{a(a.c.)} &= I_m(X - X')\cos(\omega_0 t + \lambda) \\ &= \mathrm{Re}[I_m(X - X')\,\epsilon^{j(\omega_0 t + \lambda)}] \end{aligned} \tag{11.27}$$

and

$$\begin{aligned} U_1 &= jI_{10}(X - X') \\ &= U_{10} - jX'I_{10} = U'. \end{aligned} \tag{11.28}$$

Eqn. (11.27) shows that, when the supply is disconnected, the terminal voltage of the machine drops rapidly by an amount equal to the voltage drop across the transient reactance. This result could have been deduced by the method of Section 8.6.

11.3.3 Indirect short-circuit

The analysis is simplified by changing the origin to time t_1 so that the new time is $t' = t - t_1$. The phase angle has a new value λ', which depends on t'. The complex voltage at the instant of short-circuit is

$$U_1 = jI_{10}(X - X')\epsilon^{-t/T_0'} \tag{11.29}$$

Using Eqn. (11.14) to replace I_{10}, Eqn. (11.29) becomes

$$U_1 = -jU_0'\epsilon^{j\lambda'}\epsilon^{-t'/T_0'} \tag{11.30}$$

where

$$U_0' = jI_m(X - X')\epsilon^{-t_1/T_0'} \tag{11.31}$$

U_0' is the amplitude of the open circuit voltage at the instant of short-circuit application and is also the voltage behind transient reactance.

The current before the short-circuit is zero and the Laplace transform of the current after $t' = 0$ is due to the abrupt change of voltage U_1. From Eqn. (11.30)

$$\bar{U}_1 = \frac{-jU_0'\epsilon^{j\lambda'}}{p + 1/T_0'}, \tag{11.32}$$

and from Eqns. (11.15) and (11.32)

$$\bar{I}_1 = \frac{-j\omega_0 U_0'\epsilon^{j\lambda'}}{(p + \alpha + j\omega_0)(p + 1/T_0')X(p)} \tag{11.33}$$

Solution of Eqn. (11.33) by Laplace transforms gives the current after an indirect short-circuit.

$$i_a = \frac{U_0'}{X''}\epsilon^{-t/T_a}\cos\lambda' - U_0'\left[\frac{1}{X'}\epsilon^{-t'/T'} + \left(\frac{1}{X''} - \frac{1}{X'}\right)\epsilon^{-t'/T''}\right]\cos(\omega_0 t' + \lambda'). \tag{11.34}$$

For the initial period after the short circuit, putting $\epsilon^{-t'/T'}$ and $\epsilon^{-t'/T''}$ equal to unity, the symmetrical a.c. component of short-circuit current is given by Eqn. (11.35). This result is similar to that found in Section 8.1 for the synchronous machine.

$$i_a(\text{a.c.}) = -\frac{U_0'}{X''}\cos(\omega_0 t' + \lambda'). \tag{11.35}$$

The applied voltage is given by the voltage behind transient reactance before the open circuit, multiplied by the decrement factor depending on the interval before the short-circuit. It is however, preferable to measure the value from a voltage oscillogram.

By using the equations derived above, the oscillograms obtained by a direct or indirect short-circuit can be used to determine the transient parameters, as explained in Section 8.2.

11.3.4 Fault calculations

It is clear from the foregoing that the behaviour of an induction motor under fault conditions corresponds exactly to that of a synchronous machine, except that there is no field winding. Hence the method explained on p. 191 can be applied directly to a system containing both synchronous and induction machines. Each machine is represented by the voltage behind subtransient reactance, with the subtransient reactance in series. The initial current after the fault is calculated by a straight network computation. As in a system containing only synchronous machines, an exact determination of the manner in which the current decays is only

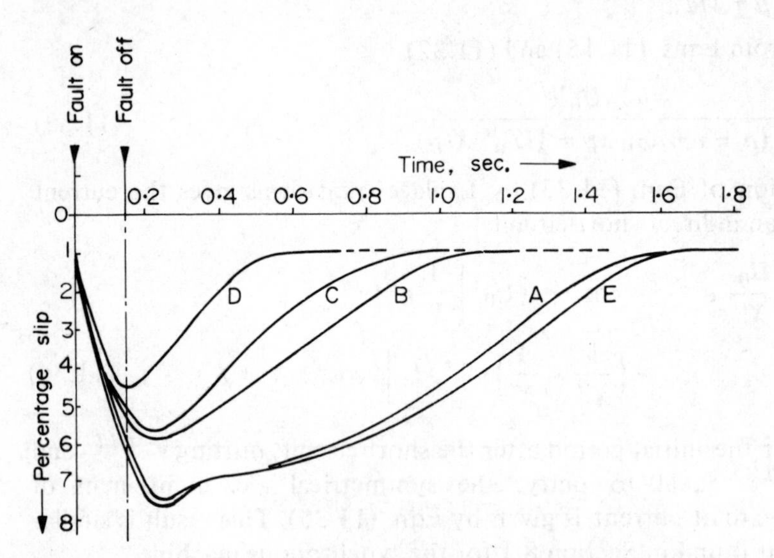

Fig. 11.3 Speed recovery of motor after temporary three-phase short-circuit.
Curves A to D were calculated as explained on p. 253
Curve E is the measured curve.

possible by a full step-by-step computation, of the kind referred to in Section 11.4.

The subtransient reactance of a deep-bar induction motor is often assumed to equal the impedance measured on a locked rotor test, for which established methods of calculation exist, using the theory of Section 10.4. The method is however not very accurate for several reasons:

(a) Since by definition the subtransient reactance is the value of $X(p)$ at infinite frequency, there is an appreciable error in taking the value at 50 Hz, especially when the effect of the deep-bars is important.
(b) Because of the approximations on which the theory is based, the parameters calculated from a steady a.c. test differ to some extent from those obtained from a short-circuit test. This has long been recognized for synchronous machines.
(c) Locked rotor tests are usually taken at reduced voltage.

Transient parameters should therefore preferably be determined by a short-circuit test if they are required for fault calculations. For design calculations they should be determined from the operational impedance locus by fitting the nearest equivalent locus applicable to a machine with two secondary windings. The matter is discussed in detail in Ref. [57].

11.4 Transient stability calculations

Consider a power system containing synchronous machines and induction motors. After a fault, the initial peak current at any point in the system is determined by a network calculation, assuming that the speed is unchanged, as indicated in the last section. During the succeeding time interval, however, the synchronous machines swing to increasing rotor angles, while the induction motors drop in speed. After the fault is cleared, the stability of the system depends on whether all the synchronous machines remain in synchronism, and on whether all the induction motors recover to the original speed just below synchronism. Because of the speed variation, the equations are non-linear and the only method available to determine the variation of angle or speed of the various machines is a numerical step-by-step computation.

For this purpose it is best to arrange the equations in state

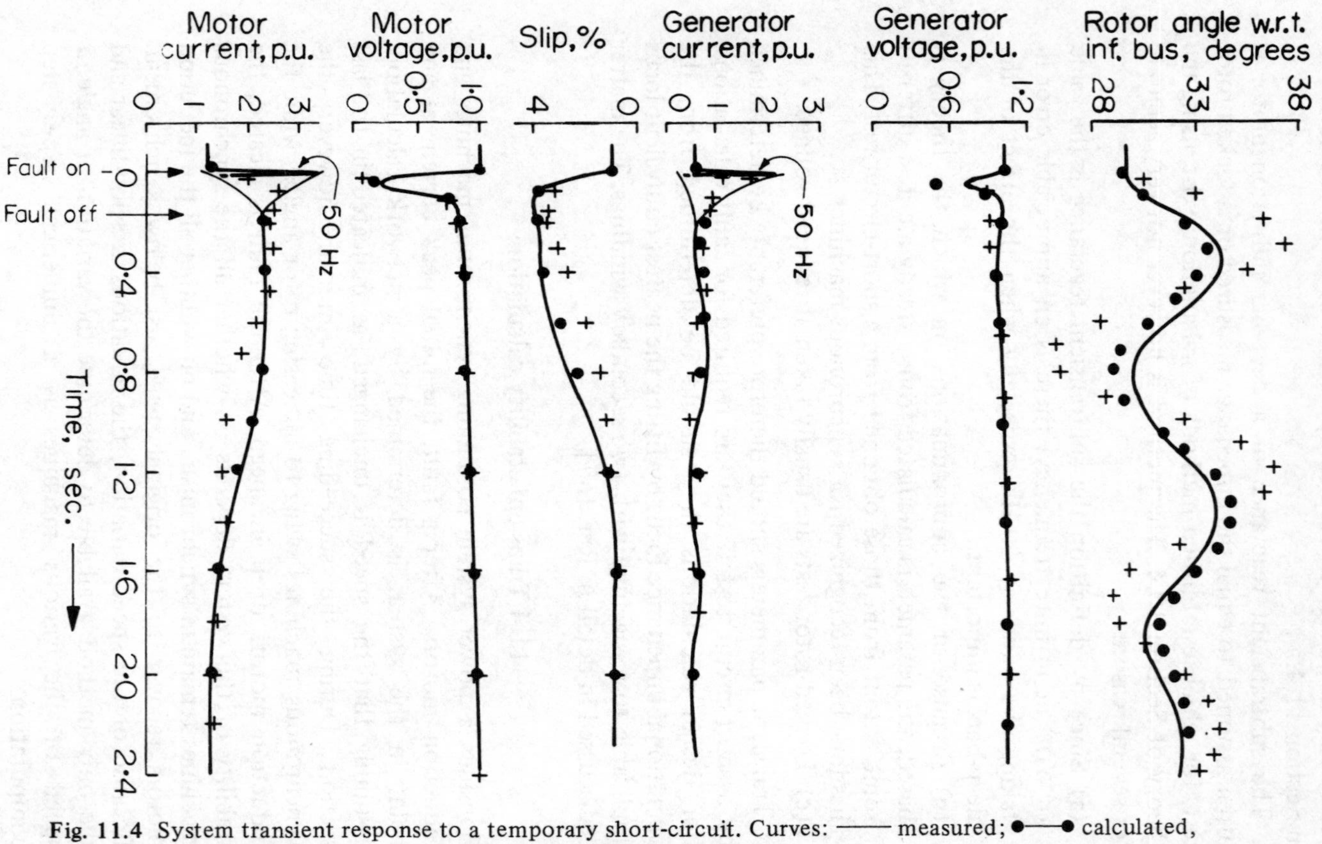

Fig. 11.4 System transient response to a temporary short-circuit. Curves: —— measured; ●—● calculated, accurate method; +—+ calculated, approximate method.

variable form. Because of the symmetry of the induction motor it is possible to halve the number of equations by using complex variables, as in the last section, but since a computer cannot directly handle complex numbers, unless the real and imaginary parts are defined individually, the equations have to be expanded again into real and imaginary parts when writing the programme, and no advantage is gained. In a multi-machine system, the equations of an induction motor are handled in almost exactly the same way as those of a synchronous machine. For a particular problem, arising in a particular system, there is always the question of how far approximations in the equations can be made. The most important such question is whether it is permissible to omit the $p\psi$ terms from Eqn. (11.2).

Fig. 11.3 shows some results obtained by test and calculation on a laboratory micro-machine. The induction motor, which was operating on load when supplied from an infinite bus with a reactance in series, was short-circuited at its terminals, followed by clearance of the fault after 0.15 seconds. The methods used to calculate the curves of Fig. 11.3 are as follows:

A. Full use of Park's equations
B. Neglecting $p\Psi$ term and replacing ω by ω_0 in Eqn. (11.7)
C. Neglecting $p\Psi$ only in Eqn. (11.7)
D. Representing the induction motor by its steady-state equivalent circuit, using the instantaneous value of slip.

In all the calculations a term was included in the torque equation to allow for the effect of load loss. If this was not done there was an appreciable error.

Fig. 11.4 shows measured and calculated curves for a model multi-machine system, in which a synchronous generator and an induction motor are connected through various transmission reactances to an infinite bus. The approximate method of calculation caused less error in the synchronous machine quantities than in those of the induction motor, presumably because the latter departs much more from synchronous speed.

It appears therefore that any method short of the full representation leads to an inaccurate result. The use of the equivalent circuit of the induction motor, which is sometimes advocated, leads to a particularly large error. The problem of induction motor transients is discussed in detail in [57].

Chapter Twelve

Application to Less Common Types of Machine

12.1 Classification in relation to the theory

It has been pointed out in Section 4.1 that the two-axis theory of a.c. machines depends on the assumption that the armature winding does not produce m.m.f. space harmonics and does not respond to space harmonics of air-gap flux density. It was shown that this is equivalent to the assumption that the inductance coefficients in the phase equations, expressed as functions of rotor position, do not contain third or higher harmonics, and also that certain relations exist between these coefficients. Where the assumption causes too much error, it is possible to use phase equations in which the higher harmonics are included. The values of the inductance coefficients can be obtained by measurement or calculation, using the principles explained in Section 1.6 to calculate the flux and hence the voltage produced by a given current in any angular position of the rotor. Although the coefficients in the equations are now functions of time, a digital computer can carry out a numerical solution.

However, the assumption is still made that any flux is proportional to the current producing it. Consequently saturation and eddy currents in the iron are not allowed for. In cases where these effects can be neglected, computations based on phase equations can give accurate results.

A classification can be made on the following lines

1. Machines for which the two-axis theory gives accurate results. All three-phase induction and synchronous machines of normal design come under this heading and hence the great majority of practical machines built at the present time. The word normal indicates that great pains are taken in the design to reduce harmonics and other subsidiary effects to a minimum and hence to comply with the assumptions of the two-axis theory.
2. Machines for which the two-axis theory can help greatly in understanding the processes involved, but cannot give high accuracy. d.c. and a.c. commutator machines are usually in this category, because of the uncertain value of any parameter which depends on commutation.
3. Problems where the two-axis equations are valid, but give no advantage over the phase equations. An induction motor supplied from a thyristor bridge comes under this heading. The accurate computation of unbalanced operation of synchronous machines is also of this type, for which the (α, β) reference frame, discussed in Section 4.1, is probably the most useful [19]. However, for steady conditions or for fault calculations a reasonably accurate approximate calculation can be made by combining the three-phase equations with symmetrical component theory.
4. Machines for which the two-axis assumptions are not justified, but saturation and eddy currents are not serious. Examples are the linear motor, in which the poles do not repeat regularly, certain change-pole motors where appreciable harmonics have to be tolerated [32], and some problems in synchronous machines described in Section 12.3. The phase equations can then be used.
5. Saturation and eddy currents always lead to reduced accuracy with either method. Where accurate transient calculations are not possible, the two-axis equations can often be used to deal with steady operation. Any machine whose main operation depends on eddy currents is an example of this condition since eddy current theory for alternating conditions is much simpler than for transients.

12.2 Application of two-axis theory

Many types of a.c. machine can be analyzed by the two-axis theory. The application to a.c. commutator machines is illustrated below by the single-phase repulsion motor. Interconnected systems are illustrated by a pair of induction motors used as power selsyns. Reluctance motors and change-pole induction motors can be analyzed by the two-axis method, except where the winding arrangement is irregular enough to invalidate the assumption about the winding harmonics. Analysis of change-pole motors under transient conditions has received little attention and it is not possible to say at what point the two-axis theory would become too inaccurate.

In the following section the bracket notation is used for matrices in order to avoid confusion with the bold letters indicating phasors.

12.2.1 Repulsion motor

Fig. 12.1 is a diagram of a repulsion motor in which the stator winding is on the direct axis, while the armature has a single pair of brushes displaced by α from the direct axis. The corresponding primitive machine is Fig. 1.5 with winding G omitted. The transformation from the (a,f) frame of reference to the (d,q,f) frame is based on the fact that the resultant m.m.f. produced by the two fictitious windings D and Q is equal to that produced by the actual armature winding A. Hence the current transformation is given by

$$\begin{bmatrix} i_f \\ i_d \\ i_q \end{bmatrix} = \begin{bmatrix} 1 & \\ & \cos\alpha \\ & \sin\alpha \end{bmatrix} \begin{bmatrix} i_f \\ i_a \end{bmatrix} \qquad (12.1)$$

or $[i] = [C]\,[i']$

The primitive impedance matrix $[Z]$ is found by omitting G from Eqn. (1.5) and putting $L_d = L_q = L_a$.

Hence

$$[Z] = \begin{bmatrix} R_f + L_{ff}p & L_{md}p & \\ L_{md}p & R_a + L_a p & L_a\omega \\ -L_{md}\omega & -L_a p & R_a + L_a p \end{bmatrix} \qquad (12.2)$$

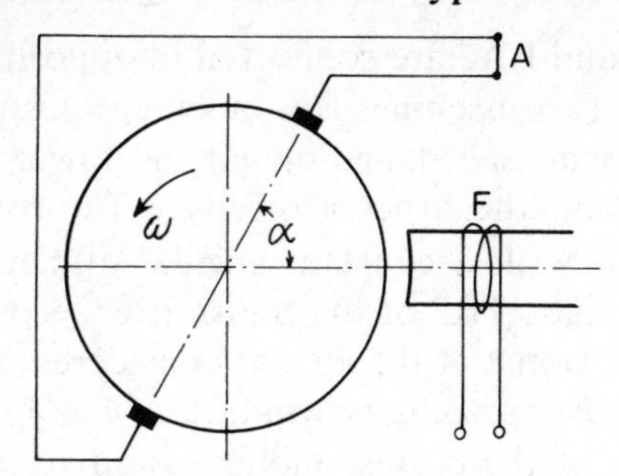

Fig. 12.1 Diagram of a repulsion motor.

The power is invariant (see Section 13.2). Hence $[Z'] = [C^T]\,[Z]\,[C]$ and the voltage equations are found to be

$$\begin{bmatrix} u_f \\ u_a \end{bmatrix} = \begin{bmatrix} R_f + L_{ff}p & L_{md}p\cos\alpha \\ L_{md}(p\cos\alpha - \omega\sin\alpha) & R_a + L_a p \end{bmatrix} \begin{bmatrix} i_f \\ i_a \end{bmatrix}$$

12.2.2 Interconnected systems

Often two or more machines are connected together in such a way that each affects the action of the other. For such combinations the system must be considered as a whole, and it is necessary to derive a combined set of equations in which the variables are all the independent currents of the complete system and all the independent speeds. The source voltages and the impressed torques corresponding to the currents and speeds also appear in the equations. The system equations can be conveniently derived by first obtaining the equations of each component separately and then combining them together by means of a connection matrix. In a system containing d.c. machines, for which the currents in the equations are the actual currents in the external circuits, the process is no more complicated than for a static network.

With a.c. machines, where the axis currents in the equations are related to the external currents by a transformation, the conditions may be more difficult, as illustrated by the example of two induction motors connected as power selsyns in Fig. 12.2a. The machines are ordinary induction motors with three-phase windings on both primary and secondary, for each of which only phase A is shown. The primary windings TA1 and RA1 in Fig. 12.2a are connected to a common supply voltage U, and the secondary windings TA2 and RA2 are connected in opposition. Under steady conditions the two machines run in synchronism with each other at exactly the same speed, and operate as power selsyns, one being a transmitter and the other a receiver. The rotors run with the same slip s and with a constant angular displacement δ between them. Thus if phase TA2 of the transmitter secondary winding has the angular position θ at the instant considered, the angle of phase RA2 of the receiver secondary winding is $(\theta - \delta)$.

The diagram of the corresponding primitive machines is shown in Fig. 12.2b, in which only the direct-axis coils T1, T2, R1, R2 are indicated. The primitive impedance matrix, treating the four coils separately, is:

$[Z] =$	$t1$	$t2$	$r1$	$r2$
$t1$	$R_{t1} + jX_{t1}$	jX_{tm}		
$t2$	jsX_{tm}	$R_{t2} + jsX_{t2}$		
$r1$			$R_{r1} + jX_{r1}$	jX_{rm}
$r2$			jsX_{rm}	$R_{r2} + jsX_{r2}$

(12.4)

where X, with appropriate suffixes, denotes the complete self-reactances and the magnetizing reactances of the two machines at supply frequency.

In order to allow for the secondary connection it is necessary to obtain a relation between the current phasors I_{t2} and I_{r2}. Because of the connection the actual secondary phase currents are equal and opposite. Hence if the windings of the two machines are identical the per-unit phase currents are related by:

$$I_{ta2} = -I_{ra2}.$$

Now for steady polyphase operation the axis currents are related by:

$$I_{td2} = jI_{tq2} = I_{t2},$$

and

$$I_{ta2} = I_{td2}\cos\theta + I_{tq2}\sin\theta$$
$$= I_{t2}\epsilon^{-j\theta}.$$

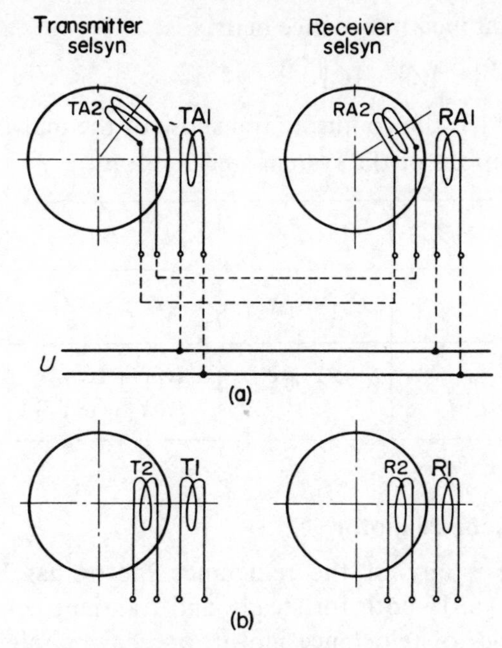

Fig. 12.2 Diagram of two power selsyns.
(a) Phase diagram.
(b) Primitive machines diagram.

Similarly

$$I_{ra\,2} = I_{r\,2}\,\epsilon^{-j(\theta - \delta)}.$$

Hence

$$I_{t\,2} = -\epsilon^{j\delta} I_{r\,2}.$$

The connection matrix is therefore the following:

$[C] =$	$t1$	$r1$	2
$t1$	1		
$t2$			1
$r1$		1	
$r2$			$-\epsilon^{-j\delta}$

(12.5)

and the combined impedance matrix is:

$$[Z'] = [C^{T*}] \cdot [Z] \cdot [C].$$

where $[C^{T*}]$ is the conjugate transpose of the matrix $[C]$.

The equations of the system are therefore:

$$\begin{bmatrix} U \\ U \\ \\ \end{bmatrix} = \begin{bmatrix} R_{t1} + jX_{t1} & & jX_{tm} \\ & R_{r1} + jX_{r1} & -jX_{rm}\epsilon^{-j\delta} \\ jsX_{tm} & -jsX_{rm}\epsilon^{j\delta} & R_{t1} + R_{r2} + js(X_{t2} + X_{r2}) \end{bmatrix} \begin{bmatrix} I_{t1} \\ I_{r1} \\ I_2 \end{bmatrix} \tag{12.6}$$

12.2.3 Reluctance motor

A thorough study of the reluctance motor has been made by Lawrenson [63], both for steady and transient conditions. There are two types of reluctance motor, one having salient poles like a normal synchronous machine and the newer *segmented type*, in which the pole pieces are isolated magnetically from each other, thereby reducing the quadrature-axis reactance.

The two-axis theory is used for computing the characteristics, but some special problems arise in determining the parameters, particularly the main axis inductances and the effective impedance of the rotor cage winding. Lawrenson gives results for steady operation, sudden load changes, run-up curves during starting and synchronizing, and asynchronous operation. The results agree quite well with measured curves, although rather less well than for large synchronous machines. The method has been valuable for determining the detailed dimensions for the best design.

A single-speed reluctance motor has a normal induction motor primary winding on the stator, for which the space harmonics are relatively small. The rotor winding consists of many circuits round the bars of the cage (see p. 75). For an accurate theory a separate damper circuit is required for each pair of bars, but for asynchronous operation at constant slip a single impedance, which varies with slip frequency, can be derived. For a transient condition, it is quite feasible to use the full equations but a shorter computation is possible using impedance parameters which are definite functions of slip, provided the reduced accuracy is acceptable.

Analysis of the magnetic circuits makes use of the concept of permeance. Although as pointed out on p. 68, the concept is not a rigorous one, it is very useful in cases where the air-gap under the poles is small, so that in this region, where most of the flux passes, the permeance is inversely proportional to the air-gap length, while in the interpolar region, where the permeance concept is less accurate, the flux is much less. In a segmented rotor, in which the magnetic potential is different in different segments, a complicated field problem has to be solved. That good results are obtained may perhaps be taken to indicate that the validity of Park's equations is influenced more by freedom from harmonics in the stator winding than by the saliency of the poles.

12.3 Application of the phase equations

Two examples where the phase equations have given accurate results, but where the results obtained by using the two-axis theory were inaccurate, are described briefly below.

Dunfield and Barton [38] calculated the wave shape of current in the armature and field windings of a synchronous machine connected to an a.c. supply. A damper winding is present on each axis and the expressions for the inductances as functions of angular position are assumed to contain Fourier terms in $\cos 3\theta$ and $\cos 4\theta$, as set out on p. 59. Two alternative methods of calculation were used.

1. The equations were expressed in state-space form and a time solution was computed by a Runge-Kutta integration.
2. A shorter computation method was obtained by assuming the armature current to contain harmonics up to the 7th, and the field winding up to the 6th. By introducing harmonic coefficients as new variables, a new set of equations with constant coefficients is derived, and the solution is computed by the shorter Newton-Raphson method.

The principal conclusion is that appreciable harmonic currents flow if the neutral of a star-connected machine is connected to the star point of the supply, but that they are small if the neutral is not connected. The computed results agreed closely with tests taken on a small salient-pole synchronous machine. Calculations based on the two-axis theory showed no harmonics in either case, and gave a satisfactory result only when the neutral was not connected.

Rajaraman and Carter [51] considered the problem of hunting in a synchronous machine without a damper winding. A similar method introducing harmonic factors was used, but the resulting equations were considerably more complicated than for the case described above. The conclusion was however similar, namely that the two-axis method was satisfactory for a three-wire connection, but that the phase equations must be used to obtain an accurate result when the neutral is connected.

12.3.1 Linear motor

The two-axis theory cannot be applied to the linear motor, because there is no periodic repetition in its construction. The motor can however be studied in principle by means of the phase equations, which are the circuit equations of a system of coils in which relative motion takes place. Each inductance coefficient is a function of the relative position of the two elements but it is not a periodic function as in a rotating machine. If there is no saturation the equations are linear with variable coefficients. Mention of such a study is made in [54] but calculations of this kind have not often been made, presumably because of the difficulty of determining and handling the large amount of data required and the fact that saturation is usually appreciable in linear motors.

Chapter Thirteen
Appendices

13.1 Representation of a.c. and transient quantities by complex numbers. The generalized phasor

Elementary treatments of the representation of a.c. quantities by phasors depend on geometrical line diagrams, and complex numbers are introduced at a later stage as a means of making calculations based on the diagrams. For a general machine theory in terms of algebraic and differential equations the phasor, which can be an important aid to calculation for certain problems, must have an algebraic definition, as in the following paragraphs.

An alternating current is fully expressed in the following equation

$$i = I_{\mathrm{m}} \cos(\omega t + \theta),$$

where

θ = the phase angle of the current,

ω = 2π times the frequency.

$$i = \mathrm{Re}\left[\sqrt{2}\left(\frac{I_{\mathrm{m}}}{\sqrt{2}}\,\epsilon^{j\theta}\right)\epsilon^{j\omega t}\right]$$

$$\delta = \mathrm{Re}[\sqrt{2}\boldsymbol{I}\epsilon^{j\omega t}] \qquad (13.1)$$

where

$$\boldsymbol{I} = \frac{I_{\mathrm{m}}}{\sqrt{2}}\,\epsilon^{j\theta} \qquad (13.2)$$

Equations (13.1) and (13.2) give the transformation in both directions between i, the instantaneous function of time and $\boldsymbol{I}$, the constant phasor or complex number.

For a system governed by a set of linear differential equations with constant coefficients, solutions, for which the variables are sinusoidal at a constant angular frequency ω, can be found by means of the following rules.

1. Change the symbols for all variables by replacing the small letters (instantaneous values) by bold capital letters (complex numbers).
2. Replace $\mathrm{d}/\mathrm{d}t$ by $j\omega$.

It should be emphasized that the phasor method, like the principle of superposition and the operational methods explained in Section 13.3, only applies if the differential equations are linear and have constant coefficients. A circuit governed by a non-linear differential equation, or one having non-linear elements, cannot have a solution in which all the voltages and currents are sinusoidal. This point brings out again the importance of the initial assumption that there is no saturation.

The use of the factor $\sqrt{2}$ conforms with the usual convention that the amplitude of the phasor is the r.m.s. value of the current. For other quantities it may be convenient to use the maximum value.

As an example, the equation of a simple inductive circuit is

$$Ri + L\frac{\mathrm{d}i}{\mathrm{d}t} = u$$

If u is an applied a.c. voltage represented by phasor $\boldsymbol{U}$, the phasor equation is

$$(R + j\omega L)\boldsymbol{I} = \boldsymbol{U}$$

The mean power in a circuit, when the current and voltage are expressed by phasors $\boldsymbol{I}$ and $\boldsymbol{U}$ is given by

$$P = \mathrm{Re}[\boldsymbol{U}^*\boldsymbol{I}] = \mathrm{Re}[\boldsymbol{U}\boldsymbol{I}^*]. \tag{13.3}$$

The generalized phasor

The transformation of Eqn. (13.1) can be applied to any current wave, varying with time in any manner. If the frequency ω is still a constant, the transformation is

$$i(t) = \mathrm{Re}\{\sqrt{2}[I_\mathrm{D}(t) + jI_\mathrm{Q}(t)]\,\mathrm{e}^{j\omega t}\} \tag{13.4}$$

The suffixes D and Q are used, because I_D and I_Q are the components of the currents in the (D,Q) reference frame, rotating at the synchronous speed determined by the supply frequency, as explained on p. 208. The complex quantity $[I_D(t)+jI_Q(t)]$ of which the components may be any possible function of t, is called a *generalized phasor.*

The concept is easily understood for the condition, commonly assumed in power-system analysis, when the axis current is a slowly varying time function and the primary current is a slowly changing supply frequency wave. In this case the generalized phasor representing the current obtained by transforming to the (D,Q) frame, is a slowly varying complex function of time. It is sometimes helpful, even when the axis current varies rapidly as in the multi-machine problem of Section 9.5, to transform the current to the (D,Q) reference frame; in other words, to represent the current by its generalized phasor. The method applies equally to voltage, or indeed to any time varying quantity.

13.2 Current and voltage transformations when power is invariant

The primitive machine and the two-axis equations derived from it can be used by means of suitable transformations to develop the theory of almost any type of rotating machine, including the most important practical machines. For the synchronous and induction machines discussed in the main part of the book, the axis equations are first solved and the transformation is used to convert the axis variables to the actual phase values. In other cases it may be better to obtain transformed equations containing the actual variables to use them for the solution.

Consider as an example that it is required to transform the variables i_1, i_2, i_3 into new variables i_1', i_2' related to the old variables by Eqn. (13.5)

$$\begin{bmatrix} i_1 \\ i_2 \\ i_3 \end{bmatrix} = \begin{bmatrix} C_{11} & C_{12} \\ C_{21} & C_{22} \\ C_{31} & C_{32} \end{bmatrix} \begin{bmatrix} i_1' \\ i_2' \end{bmatrix} \qquad (13.5)$$

or

$$\boldsymbol{i} = \boldsymbol{C}\boldsymbol{i}' \tag{13.6}$$

Assume that the voltage transformation is

$$\boldsymbol{u}' = \boldsymbol{B}\boldsymbol{u} \tag{13.7}$$

where $\boldsymbol{B} = \begin{array}{c|c|c} B_{11} & B_{12} & B_{13} \\ \hline B_{21} & B_{22} & B_{23} \end{array}$

When $\boldsymbol{u}$ and $\boldsymbol{i}$ are expressed in actual units (volts and amperes), the power is *invariant,* that is, the same value in watts is obtained with either set of variables. Hence

$$P = u_1 i_1 + u_2 i_2 + u_3 i_3 = u_1' i_1' + u_2' i_2' \tag{13.8}$$

Substituting i_1, i_2, i_3 from Eqn. (13.5) and u_1', u_2' from Eqn. (13.7),

$$C_{11}u_1 i_1' + C_{12}u_1 i_2' + C_{21}u_2 i_1' + C_{22}u_2 i_2' + C_{31}u_3 i_1' + C_{32}u_3 i_2'$$

$$\equiv B_{11}u_1 i_1' + B_{12}u_2 i_1' + B_{13}u_3 i_1' + B_{21}u_1 i_2' + B_{22}u_2 i_2' + B_{23}u_3 i_2'$$

This identity is true for all values of u_1, u_2, u_3, i_1', i_2'. Hence

$$\boldsymbol{B} = \boldsymbol{C}^{\mathrm{T}} \tag{13.9}$$

where $\boldsymbol{C}^{\mathrm{T}}$ is the *transpose* of the matrix $\boldsymbol{C}$.

Hence if it is known that the power is invariant, it is only necessary to determine the current transformation matrix $\boldsymbol{C}$. The voltage transformation matrix is $\boldsymbol{C}^{\mathrm{T}}$, so that $\boldsymbol{u}' = \boldsymbol{z}'\boldsymbol{i}'$. The transformed impedance matrix is given by

$$\boldsymbol{Z}' = \boldsymbol{C}^{\mathrm{T}}\boldsymbol{Z}\boldsymbol{C} \tag{13.10}$$

Note that the current transformation is from new to old, while the voltage transformation is from old to new.

When a particular per-unit system is used, the power may not necessarily be invariant in the mathematical sense. An important example is the Park transformation Eqn. (4.4). It has been proposed that the transformation should be changed so as to introduce a factor $\sqrt{\frac{3}{2}}$ in order to make the power invariant. From a practical point of view this would be undesirable and would introduce an unnecessary complication, whereas the fact that P is not invariant causes no trouble.

13.3 Operational methods

Earlier work on the transient performance of electrical machines uses the Heaviside notation, in which $\mathrm{d}/\mathrm{d}t$ is replaced by p in the differential equations. The abbreviation is convenient, whether or not the equations are linear. For linear equations, the Heaviside method provides also a means of obtaining a solution by manipulating them as though p were an ordinary algebraic quantity, and then applying certain well-defined rules, expressed as *standard forms*, to determine the variables as functions of time. p is however not an ordinary quantity but an *operator* having special properties, and the solution, which is not mathematically rigorous, is referred to as an *operational solution*. It applies directly only to systems for which the initial values of the variables are zero, but it can be applied when the initial values are not zero by using the principle of superposition. In the early development of control system theory, many of the important techniques (Nyquist, Bode, etc.) made use of the p equations without undertaking the labour of computing the complete solution. When expressed in the Heaviside notation, the same symbols are used for the variables in the transformed equations as in the original differential equations. For example, a transfer function relating two variables is often stated as a ratio of the time variables, or an operational equivalent circuit is drawn, indicating the voltages and currents by their time symbols, but including *operational reactances*, of the type pL.

The Laplace transform method makes possible a rigorous solution of the differential equations by transforming any time variable into a new variable in the p-domain. (Most recent texts on control system theory use the symbol s, but p is used here, because s is a well-established symbol for *slip* in the theory of electrical machines.) In the Laplace transform theory, p is now an ordinary algebraic variable (often complex) and the transformed main variable should be denoted by a different symbol from the time variable, for example, the transform of $i(t)$ is denoted by $\bar{i}(p)$. For subsequent manipulation of the p equation in deriving transfer

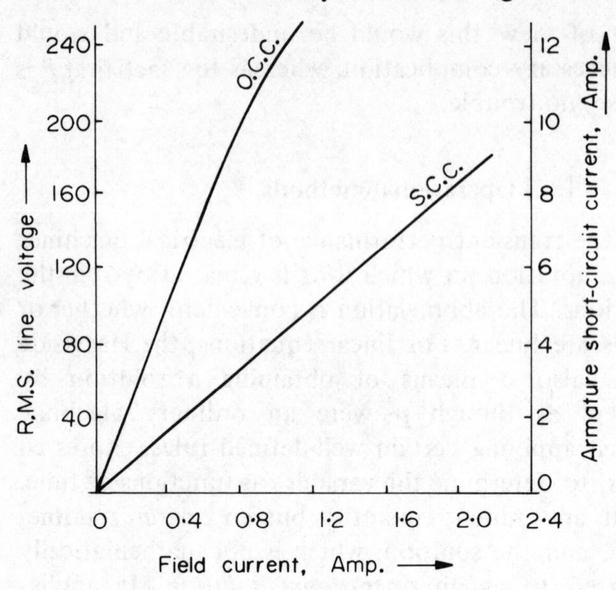

Fig. 13.1 Open-circuit and short-circuit characteristic.

functions or equivalent circuits, it is convenient to revert to the original symbol for the time variable (without the bar), when p has to be again regarded as an operator.

The treatment in the present book is a combination of the two methods. The original formulation of the differential equations is made less cumbersome by making a straight substitution of p for d/dt but analytical solutions are only obtained when the equations are linear. The basic solution only applies to cases where the variables are initially zero, but solutions when they are not zero are always possible by the principle of superposition, because the superimposed variables are necessarily initially zero. Thus the Laplace transform method, applied to an initially dead system, provides a rigorous verification of Heaviside's method. In the text, symbols with a bar above are used in the course of a solution to indicate that the method is a rigorous application of the Laplace transform method. In the subsequent application of the transformed equations to transfer functions and equivalent circuits the bar is dropped, so that p again becomes effectively an *operator*.

The following transforms are used for solutions given in the text.

Table 13.1 Table of inverse Laplace transforms.

Used on page	$f(p)$	*Inverse of* $f(p)$	
40	$\dfrac{p+c}{p(p+a)(p+b)}$	$\dfrac{c}{ab}+\dfrac{(c-a)\epsilon^{-at}}{a(a-b)}+\dfrac{(c-b)\epsilon^{-bt}}{b(b-a)}$	(13.11)
163 164	$\dfrac{\omega_0{}^2}{(p+a)(p^2+2\alpha p+\omega_0{}^2)}$	$\epsilon^{-at}-\epsilon^{-\alpha t}.\cos\omega_0 t$ assuming that $a, \alpha \ll \omega_0$	(13.12)
163	$\dfrac{\omega_0 p}{(p+a)(p^2+2\alpha p+\omega_0{}^2)}$	$\epsilon^{-\alpha t}.\sin\omega_0 t$ assuming that $a, \alpha \ll \omega_0$	(13.13)
246	$\dfrac{\omega_0}{p(p+a)(p+\alpha+j\omega_0)}$	$j[\epsilon^{-(\alpha+j\omega_0)t}-\epsilon^{-at}]$ assuming that $\alpha, a \ll \omega_0$	(13.14)

13.4 The per-unit system [52]

13.4.1 The transformer

The following explanation of the per-unit system as applied to a transformer amplifies that given in Section 1.2.

Let $u_1{}^a$, $u_2{}^a$, $i_1{}^a$, $i_2{}^a$ be the *actual* values of variables in volts and amperes; let $U_1{}^b$, $U_2{}^b$, $I_1{}^b$, $I_2{}^b$ be the respective *base* values, also in volts and amperes. The *per-unit* or *normalized* values are then

$$u_1{}^p = \frac{u_1{}^a}{U_1{}^b}, \qquad i_2{}^p = \frac{i_2{}^a}{I_2{}^b}, \qquad \text{etc.,}$$

In Section 1.2 the secondary base values are chosen such that

$$\left.\begin{aligned} U_2{}^b &= NU_1{}^b \\ I_2{}^b &= I_1{}^b/N \end{aligned}\right\} \qquad (13.15)$$

where N is the ratio of secondary turns N_2 to primary turns N_1. This particular choice of relationship between primary and secondary base values ensures that base power on the primary side equals base power on the secondary side so that

$$P_1{}^b = U_1{}^b I_1{}^b = P_2{}^b = U_2{}^b I_2{}^b \qquad (13.16)$$

Since the base value of time is one second (p. 12), self-inductance in either circuit has the same base value as impedance. Thus the base values of impedance, self-inductance and resistance are

$$\left.\begin{aligned} R_1{}^{b} &= L_{11}{}^{b} = Z_1{}^{b} = U_1{}^{b}/I_1{}^{b} \\ R_2{}^{b} &= L_{22}{}^{b} = Z_2{}^{b} = U_2{}^{b}/I_2{}^{b} \end{aligned}\right\} \tag{13.17}$$

The base mutual inductance, using Eqns. (13.15) and (13.17), is

$$L_{12}{}^{b} = L_{21}{}^{b} = U_2{}^{b}/I_1{}^{b} = U_1{}^{b}/I_2{}^{b} = NL_{11}{}^{b} = L_{22}{}^{b}/N \tag{13.18}$$

With current flowing in both windings of the transformer in Fig. 1.6, the expression for the impressed voltage in coil 1 is

$$u_1{}^{a} = \left(R_1{}^{a} + L_{11}{}^{a}\frac{\mathrm{d}}{\mathrm{d}t}\right) i_1{}^{a} + L_{12}{}^{a}\frac{\mathrm{d}i_2{}^{a}}{\mathrm{d}t} \tag{13.19}$$

Eqn. (13.19) is normalized by dividing each side by $U_1{}^{b}$ and using Eqns. (13.17) and (13.18) as follows

$$\begin{aligned} u_1{}^{p} = \frac{u_1{}^{a}}{U_1{}^{b}} &= \left[\frac{R_1{}^{a}I_1{}^{b}}{U_1{}^{b}} + \frac{L_{11}{}^{a}I_1{}^{b}}{U_1{}^{b}}\cdot\frac{\mathrm{d}}{\mathrm{d}t}\right]\frac{i_1{}^{a}}{I_1{}^{b}} + \frac{L_{12}{}^{a}I_2{}^{b}}{U_1{}^{b}}\cdot\frac{\mathrm{d}}{\mathrm{d}t}\left[\frac{i_2{}^{a}}{I_2{}^{b}}\right] \\ &= \left(R_1{}^{p} + L_{11}{}^{p}\frac{\mathrm{d}}{\mathrm{d}t}\right) i_1{}^{p} + L_{12}{}^{p}\frac{\mathrm{d}i_2{}^{p}}{\mathrm{d}t} \end{aligned} \tag{13.20}$$

Normalization of the secondary equation is carried out in a similar manner.

The mutual flux Φ^{a} corresponding to m.m.f. F^{a} is

$$\Phi^{a} = P_{m}F^{a} = P_{m}(N_1 i_1{}^{a} + N_2 i_2{}^{a}) \tag{13.21}$$

where P_{m} is the permeance of the path followed by the mutual flux.

Mutual inductance $L_{12}{}^{a}$ is defined as the flux linkage with the second coil per ampere in the first coil, with $i_2 = 0$ and similarly for L_{21}. Hence taking each term of Eqn. (13.21) separately

$$L_{12}{}^{a} = L_{21}{}^{a} = P_{m}N_1 N_2 \tag{13.22}$$

Substitution of Eqn. (13.22) in Eqn. (13.21) yields

$$\Phi^{a} = L_{12}{}^{a}i_1{}^{a}/N_2 + L_{12}{}^{a}i_2{}^{a}/N_1 \tag{13.23}$$

The base value of flux is defined as

$$\Phi^{b} = L_{11}{}^{b} I_1{}^{b}/N_1 = L_{22}{}^{b} I_2{}^{b}/N_2 \tag{13.24}$$

Combining Eqns. (13.23) and (13.24), the normalized flux is

$$\Phi^{p} = L_{12}{}^{p} i_1{}^{p} + L_{12}{}^{p} i_2{}^{p} \tag{13.25}$$

To avoid complication the superscript p is not used to distinguish per-unit quantities in the main text. Where the distinction is not obvious the text indicates whether actual or per-unit quantities are being used.

Similar considerations to those discussed above apply to the induction motor, which is often called a *generalized transformer.* It may be noted that the use of the per-unit system has the same effect of eliminating N from the equations as the use of *referred* quantities in the normal text-book treatment.

13.4.2 The synchronous machine

Many of the points raised in connection with the transformer apply also to the synchronous machine, but special consideration is needed to determine the base values of field voltage and current.

The parameters of the machine are defined by the equations, which can be used to calculate the base values for the field winding from the open-circuit and short-circuit characteristics, either measured or calculated. From Eqns. (4.27), in which all quantities are per-unit, the direct-axis equations for the open and short-circuit conditions are obtained by putting $p = j\omega$, and $\omega = \omega_0$. The armature resistance is neglected.

$$U_{d0}{}^{p} = jX_{md}{}^{p} I_{f0}{}^{p} \tag{13.26}$$

$$0 = jX_{md}{}^{p} I_{fs}{}^{p} + j(X_{md}{}^{p} + X_{a}{}^{p}) I_{ds}{}^{p} \tag{13.27}$$

The axis base values are $I_d{}^{b} = \frac{3}{2} I_a{}^{b}$ and $U_d{}^{b} = U_a{}^{b}$, as explained on p. 63. If values of $U_{a0}{}^{a}$ and $I_{as}{}^{a}$ are read from the open and short-circuit curves at the same field current, the reactance is numerically

$$(X_{md}{}^{p} + X_{a}{}^{p}) = \frac{U_{d0}{}^{p}}{I_{ds}{}^{p}} = \frac{U_{a0}{}^{p}}{I_{as}{}^{p}} = \frac{U_{a0}{}^{a}}{U_{a}{}^{b}} \cdot \frac{I_{a}{}^{b}}{I_{as}{}^{a}} \tag{13.28}$$

since, as explained on p. 134, the axis voltage and current are each

$\sqrt{2}$ times the real component of the phasor representing the phase voltage and current.

The base field current is therefore, using Eqn. (13.26),

$$I_f{}^b = \frac{I_{f0}{}^a}{I_{f0}{}^p} = \frac{X_{md}{}^p I_{f0}{}^a}{U_{d0}{}^p} = \frac{X_{md}{}^p I_{f0}{}^a}{\sqrt{2}\, U_{a0}{}^p}. \qquad (13.29)$$

Since the base power has the same value in the axis and field circuits,

$$U_f{}^b = \frac{U_d{}^b I_d{}^b}{I_f{}^b} = \frac{3}{2} \cdot \frac{U_a{}^b I_a{}^b}{I_f{}^b}. \qquad (13.30)$$

Consider the case of a 3 kW, 220 V, 3-phase, star-connected synchronous machine, of which the open-circuit and short-circuit characteristics appear in Fig. 13.1. The armature base values are chosen as

$$U_a{}^b = 127 \text{ V}, I_a{}^b = 7.87 \text{ A}.$$

The open circuit (neglecting saturation) and short-circuit curves at $I_f{}^a = 0.86$ A give

$$U_{a0}{}^a = 127 \text{ V}, I_{as}{}^a = 3.60 \text{ A}, U_{a0}{}^p = 1 \text{ p.u}, I_{as}{}^p = 0.457 \text{ p.u.}$$

From Eqn. (13.28)

$$(X_{md}{}^p + X_a{}^p) = \frac{1}{0.457} = 2.19 \text{ p.u.}$$

Taking $X_a{}^p = 0.12$ p.u., the mutual impedance is $X_{md}{}^p = 2.07$ p.u.

Hence, using Eqns. (13.29) and (13.30), the base values of field current and voltage are

$$I_f{}^b = \frac{(2.07)(0.86)}{\sqrt{2}} = 1.26 \text{ A}$$

$$U_f{}^b = \frac{3}{2} \cdot \frac{(127)(7.87)}{1.35} = 1190 \text{ V}.$$

References

* An item marked with an asterisk is one of several papers on the same subject by the same author. Other references can be found in the bibliography of the paper.

Ref. No.

[1] **Doherty, R. E.** (1923). 'A Simplified method of Analysing Short-circuit Problems', *Trans. A.I.E.E.*, **42**, p. 841.

[2] **West, H. R.** (1926), 'Cross-field Theory of Alternating Current Machines', *Trans. A.I.E.E.*, **45**, p. 466.

[3] **Park, R. H.** (1928), 'Definition of an Ideal Synchronous Machine', *G.E. Review*, **31**, p. 332.

[4] **Park, R. H.** (1929), 'Two-reaction Theory of Synchronous Machines', *Trans. A.I.E.E.*, **48**, p. 716.

[5] ***Doherty, R. E.** and **Nickle, C. A.** (1930), 'Synchronous Machines I to V. *Trans. A.I.E.E.*, **49**, p. 700.

[6] **Park, R. H.** (1933), 'Two-reaction Theory of Synchronous Machines. II', *Trans. A.I.E.E.*, **52**, p. 352.

[7] **Kingsley, C.**, (1935), 'Saturated Synchronous Reactance', *Trans. A.I.E.E.*, **54**, p. 300.

[8] ***Kron, G.** (1942) 'Application of Tensors to the Analysis of Rotating Electrical Machinery', *G.E. Review.*

[9] **Szwander, W.** (1944), 'Fundamental Characteristics of Synchronous Turbo-generators', *Jour. I.E.E.*, **91**, Part II, p. 185.

[10] **Rankin, A. W.** (1945), 'The Direct and Quadrature-Axis Equivalent Circuits of the Synchronous Machine', *Trans. A.I.E.E.*, **64**, p. 861.

[11] **Crary, S. B.** (1947) *Vol. 2. 'Power System Stability Transient Stability'*, John Wiley, London.

[12] ***Gibbs, W. J.** (1948), 'Induction and Synchronous Motors with Unlaminated Rotors', *J.I.E.E.*, **95**, Part II, p. 411.

[13] **Linville, T. M.** and **Ward, H. C.** (1949), 'Solid Short Circuit of d.c. Motors and Generators', *Trans. A.I.E.E.*, **68**, p. 119.

[14] **Duesterhofft, W. C.** (1949), 'The Negative Sequence Reactances of an Ideal Synchronous Machine', *Trans. A.I.E.E.*, **68**, p. 510.

[15] **Adkins, B.** (1951) 'Transient Theory of Synchronous Generators Connected to Power Systems', *J.I.E.E.*, **98**, p. 510.

[16] **Concordia, C.** (1951), *Synchronous Machines,* John Wiley, London.

[17] **Laible, Th.** (1952), *Die Theorie der Synchronmaschine in Nichtstationarem Betrieb.* Springer.

[18] **Walker, J. H.** (1953), 'Operating Characteristics of Salient-pole Machines', *Proc. I.E.E.* Part II, **100**, p. 13.

[19] **Ching, Y. K.** and **Adkins, B.** (1954), 'Transient Theory of Synchronous Generators under Unbalanced Conditions', *Jour. I.E.E.,* **101,** Part IV, p. 106.

[20] **Adkins, B.** (1957), *The General Theory of Electrical Machines,* Chapman and Hall.

[21] **Dalal, M. K.** (1957), *The two Reactions of a Synchronous Machine.* M.Sc. Thesis, London University.

[22] **Say, M. G.** (1958), *The Performance and Design of Alternating Current Machines,* Pitman.

[23] **White, D. C.** and **Woodson, H. H.** (1959), *Electromechanical Energy Conversion* John Wiley.

[24] **Mehta, D. B.** and **Adkins, B.** (1960), 'Transient Torque and Load Angle of a Synchronous Generator following several Types of System Disturbance', *Proc. I.E.E.,* **107**, Part A, p. 61.

[25] **Concordia, C.** (1960), 'Synchronous Machine with Solid Cylindrical Rotor – Part II, *Trans. A.I.E.E.,* **78**, Part III.

[26] **Carter, G. W.** *et al.* (1961), 'The Inductance Coefficients of a Salient-pole Alternator in Relation to the Two-axis Theory', *Proc. I.E.E.,* **108, Part A, p. 263.**

[27] **Ralston, A.** and **Wilf, H. S.** (1962), *Mathematical Methods for Digital Computers,* John Wiley.

[28] **Bharali, P.** and **Adkins, B.** (1963), 'Operational Impedances of Turbo-generators with Solid Rotors', *Proc. I.E.E.,* **110**, p. 2185.

[29] **Shackshaft, G.** (1963), 'A General Purpose Turbo-alternator Model', *Proc. I.E.E.,* **110**, p. 703.

[30] **Alger, P. L.** (1965), *The Nature of Polyphase Induction Machines,* Gordon and Breach, New York.

[31] **Humpage, W. D.** and **Stott, B.** (1965), 'Predictor-corrector Methods of Numerical Integration in Digital-computer Analyses of Power-system Transient Stability, *Proc. I.E.E.,* **112**, p. 1557.

[32] ***Rawcliffe, G. H.** and **Fong, W.** (1965), 'Two-speed Induction Motors with Fractional-slot Windings' *Proc. I.E.E.,* **112**, p. 1899.

[33] **Jacovides, L. J.** and **Adkins, B.** (1966), 'Effect of Excitation Regulation on Synchronous Machine Stability', *Proc. I.E.E.,* **113**, p. 1021.

[34] **Jones, C. V.** (1967), *The Unified Theory of Electrical Machines,* Butterworths.

[35] **Elgerd, O. I.** (1967), *Control System Theory,* McGraw-Hill.

[36] **Humpage, W. D.** and **Saha, T. N.** (1967), 'Digital Computer Methods in Dynamic Response Analysis of Turbo-generator units', *Proc. I.E.E.,* **114**, p. 1115.

[37] **Smith, I. R.** and **Sriharan, S.** (1967), 'Induction Motor Reswitching Transients', *Proc. I.E.E.,* **114**, p. 503.

[38] **Dunfield, J. C.** and **Barton, T. H.** (1967), 'Effect of m.m.f. and Permeance Harmonics in Electrical Machines, with Special Reference to a Synchronous Machine', *Proc. I.E.E.,* **114**, p. 1443.

[39] BS 4296/1928. *Methods of Test for Determining Synchronous Machine Quantities.*

[40] **Prabhashankar, K.** and **Janischewsyj, W.** (1968), 'Digital Simulation of Multi-machine Power and Systems for Stability Studies', *Trans. I.E.E.E.,* PAS-87, p. 73.

[41] **Widger, G. F. T.** and **Adkins, B.** (1968), 'Starting Performance of Synchronous Motors with Solid Salient Poles', *Proc. I.E.E.,* **115**, p. 1471.

[42] **Canay, I. M.** (1969), 'Causes of Descrepancies on Calculation of Rotor Quantities and Exact Equivalent Diagrams of the Synchronous Machine', *Trans, I.E.E.E.,* PAS-88, p. 1114.

[43] **Soper, J. A.** and **Fagg, A. R.** (1969), 'Divided-winding-rotor Synchronous Generator', *Proc. I.E.E.,* **116,** p. 113.

[44] **Kapoor, S. C.** *et al.* (1969) 'Improvement of Alternator Stability by Controlled Quadrature Excitation', *Proc. I.E.E.,* **116**, p. 771.

[45] **Gupta, S. C.** and **Hasdorf, L.** (1970), *Fundamentals of Automatic Control.* John Wiley.

[46] **Speedy, C. B.,** *et al.* (1970), *Control Theory: Identification and Optimal Control.* Oliver and Boyd, Edinburgh.

[47] **Harley, R. G.** and **Adkins, B.** (1970), 'Stability of Synchronous Machine with Divided Winding Rotor', *Proc. I.E.E.,* **117**, p. 933.

[48] **Willems, J. L.** and **Willems, J. C.** (1970), 'The Application of Liapunov Methods to the Computation of Transient Stability Regions for Multi-machine Power Systems', *Trans I.E.E.E.,* PAS-89, p. 795.

[49] **Harley, R. G.** and **Adkins, B.** (1970), 'Calculation of the Angular Back Swing following a Short Circuit of a Loaded Alternator', *Proc. I.E.E.,* **117**, p. 377.

[50] **Shackshaft, G.** (1970), 'Effect of Oscillatory Torques on the Movement of Generator Rotors', *Proc. I.E.E.,* **117**, p. 1969.

[51] **Rajamaran, K. C.** and **Carter, G. W.** (1970), 'Effect of Harmonics on Hunting of Synchronous Machines', *Proc. I.E.E.,* **117**, p. 1143.

[52] **Harris, M. R., Lawrenson, P. J.** and **Stephenson, J. M.** (1970), *Per-unit Systems,* C.U.P.

[53] ***Chari, M. V.** and **Silvester, P.** (1971), 'Analysis of Turbo-generator Magnetic Fields by Finite Elements', *Trans. I.E.E.E.,* PAS-90, p. 454.

[54] ***Laithwaite, E. R.** *et al.* (1971), 'Linear Motors with Transverse Flux', *Proc. I.E.E.,* **118**, p. 1761.

[55] **Iyer, S. N.** and **Cory, B. J.** (1971), 'Optimization of Turbo-generator Transient Performance by Differential Dynamic Programming', *Trans. I.E.E.E.,* PAS-90, p. 2149.

[56] **Kalsi, S. S.** and **Adkins, B.** (1971), 'Transient Stability of Power Systems Containing both Synchronous and Induction Machines', *Proc. I.E.E.,* **118**, p. 1467.

[57] **Kalsi, S. S.** *et al.* (1971), 'Calculation of System Fault Currents due to Induction Motors', *Proc. I.E.E.,* **118**, p. 201.

[58] **Hammond, P.** (1971), '*Applied Electromagnetism*', Pergamon Press, London.

[59] **Outhred, H. R.** and **Evans, F. J.** (1972), 'A Model Reference Adaptive Controller for Turbo-alternators in Large Systems', *Fourth Power Systems Computation Conference,* Grenoble.

[60] **Mason, T. H.** *et al.* (1972), 'Asynchronous Operation of Turbo-generators', *C.I.G.R.E.*, Paper 11-02.

[61] **Shackshaft, G.** and**Neilson, R.** (1972), 'Results of Stability Tests on an Under-excited 120 MW Generator', *Proc. I.E.E.*, **119**, p. 175.

[62] ***Fuchs, E. F.** and **Erdelyi, E. A.** (1972), 'Determination of Water-wheel Alternator Steady-state Reactances from Flux Plots', *Trans. I.E.E.E.*, PAS-91, p. 1795.

[63] ***Lawrenson, P. J.** and **Mathur, R. M.** (1972), 'Asynchronous Performance of Reluctance Machines allowing for Irregular Distribution of Rotor Conductors', *Proc. I.E.E.* **119**, p. 318.

[64] **Raman, S.** and **Kapoor, S. C.** (1972), 'Synthesis of Optimal Regulators for a Synchronous Machine', *Proc. I.E.E.*, **119**, pp. 1383, 1391.

[65] **Humpage, W. D.** (Sept., 1973), 'Numerical Integration', *U.M.I.S.T.* Symposium on Power System Dynamics.

[66] **Canay, M.** (1972), 'Experimentelle Ermittlung der Ersatzschemata und der Parameter einer Idealisierten Synchronmaschine', *Bulletin des Schweizerischen Elektrotechnischen Vereins,* **63**, p. 1137.

[67] **Takeda, Y.** and **Adkins, B.** (1974), 'Determination of Synchronous Machine Parameters Allowing for Unequal Mutual Inductances', *Proc. I.E.E.*, **121**, p. 1501.

[68] **Dineley, J. L.** and **Fenwick, P. J.** (1974), 'The Effect of Prime-mover and Excitation Control on The Stability of Large Steam Turbine Generators', *Trans. I.E.E.E.*, PAS-93, p. 1613.

[69] **Rao, K. V. N.** *et al.* (1974), 'Peak Inverse Voltages in the Rectifier Excitation Systems of Synchronous Machines', *Trans I.E.E.E.*, PAS-93, p. 1685.

[70] **Faruqi, F. A.** (1973), *Optimal Feed-back Design for Transient Stability of Multi-machine Power Systems,* Ph.D. Thesis, London University.

[71] **Hannalla, A. Y.** and **Macdonald, D. C.** (1974), *Numerical Analysis of the Transient Field Problems in Electrical Machines,* Internal Report, Imperial College.

Index